Round Shot and Rammers

Ist Company Maryland Mattrosses 1776

by **HAROLD L. PETERSON**

Round Shot and Rammers

Illustrated by

Peter F. Copeland

Donald W. Holst

Robert L. Klinger

STACKPOLE BOOKS

Ad

JAMES C. HAZLETT, M.D.

atque

EMANUEL R. LEWIS, Ph.D.

Studiosissimi bellicorum tormentorum

ROUND SHOT AND RAMMERS

Published by
STACKPOLE BOOKS
Cameron and Kelker Streets
Harrisburg, Pa. 17105

Standard Book Number 8117-1501-9
Library of Congress Catalog Card Number 69-16153
Printed in U.S.A.

Preface

Artillery lends dignity to what might otherwise be a vulgar brawl." So runs the old saying, and few cannoneers will dispute it. Less biased observers may not agree entirely, but even they admit that artillery has played a major role in America's military history. For four centuries cannon have defended American positions, destroyed enemy strongholds and supported attacks by infantry and cavalry. Sometimes their role has been minor; often it has been decisive.

Men served these cannon, and their role should not be forgotten. Yet it was the cannon themselves that determined much of what took place. No crew could outperform its gun's capabilities. The individual cannon's speed, accuracy, and strength were the key factors that controlled its employment—and predicted its success or its failure. To understand an action that involved artillery the historian must know the guns themselves. Only then can he judge the competence of the artillerymen or the wisdom of the commanders.

But the cannon was more than just a tool. For centuries it was also a symbol. Veteran Confederate artillerist, Maj. Robert Stiles summed it up for many an old gunner: "The gun is the rallying point of the detachment, its point of honor, its flag, its banner. It is that to which men look, by which they stand, with and for which they fight, by and for which they fall. As long as the gun is theirs, they are unconquered, victorious; when the gun is lost, all is lost." And this applied to the supporting infantry as well as to the cannoneers themselves.

When the days of formal field battles passed, the cannon lost some of its symbolism. It became just another piece of equipment, a machine that was less expensive than some, and its rusting carcass could be left on beaches or in jungles, one more expendable item in a modern war. But this is a recent concept. Never was it true of a muzzle-loading piece. Those cannon were both treasures and symbols, and armies were honored or disgraced according to whether they held or lost their guns.

The present volume offers a brief introduction to those heady days of artillery. It is not a detailed technical treatise on stresses, design, theory, minor architectural variations, or other minutiae dear to the advanced artillery buff who specializes in a given period, type, or model. Rather it seeks to offer a general background to the military historian, a start to the artillery enthusiast, and a sense of his historic heritage to the modern artilleryman. Research for this book has stretched through twenty-five years of reading old manuals, diaries, and journals, in searching through manuscripts and drawings in the National Archives, in studying the surviving cannon themselves, even in training modern crews to fire muzzle-loading pieces in the proper manner and participating in the drill to get the proper feeling for the action. To document the necessarily broad statements in such a book as this would be pedantic in the extreme, but a bibliography of the more important published sources is appended for those who wish to know more about specific pieces or periods.

During this quarter-century of preparation I have had the unselfish help of many individuals. It would be impossible to mention all who have guided me to surviving cannon, called some salient fact to my attention, or corrected some misunderstanding. I would, however, like to single out a few who have been especially generous and extend them my thanks.

First of all to the artists, Peter Copeland, Donald W. Holst, and Robert L. Klinger, who worked far beyond any reasonable expectation for reward and who cheerfully put up with numerous niggling corrections and alterations.

Next, to those serious students of muzzle-loading artillery—Dr. James C. Hazlett, Dr. Emanuel R. Lewis, Edwin C. Rich, William E. Meuse, and CWO Sydney C. Kerksis as well as the late Cols. Harry C. Larter and Cary S. Tucker—who made facts available from their own research, discussed theories at length and at all hours, and some of whom also read and criticized this work in manuscript.

To Samuel E. Stetson, who helped mightily with the captions for the plates.

And finally to my wife, Dorothy, who typed the manuscript and was her usual patient self through the emotional periods of book production.

Arlington, Virginia   HAROLD L. PETERSON

Contents

Artillery in the colonies—guns carried by the Spanish explorers / cannon in French and Spanish forts / cannon in the English and Dutch colonies / artillery traditions of the Massachusetts Bay colony / limited uses of artillery

How guns were classified

Leather guns—an experiment in mobility

Swivel guns—for rapid fire at close range

Carriages—evolution of the carriage from the stock trail type to the flask carriage and the invention of the limber / fortress carriages / carriage woods, colors, and dimensions

Ammunition in the colonial period—gunpowder ingredients and the graining process / cartridges and fixed ammunition for rapid firing / kinds of shot—langrage, cannister, grape, bar, cross bar, jointed, expanding, and chain / shells and incendiaries—bombs, carcasses, and spike shot

Duties of the artillerist—loading and firing—how the gunner used his match, linstock, sponge, powder ladle, rammer, and other implements / aiming—how the gunner used his level, quadrant, calipers, and other instruments to achieve a limited degree of accuracy

Artillery in siege warfare

Artillery in the field

British artillery—the "rose and crown" guns of the 18th century / Gen. Armstrong's standardization and John Muller's recommendations / the shortening of cannon / how to date British guns / mortars, howitzers, and petards / gun markings / gun metals / howitzer, galloper, and fortress carriages / mortar beds / "monkey tails" and pole tillers for aiming swivel guns / rolling stock

French artillery—improvements of the Vallière System—standardization of calibers, lengths, proportions, and weights; functional decorations for identification of each type of piece / adoption of the Swedish 4-pounder for field use / improvements of the Gribeauval System—different designs for field, siege, garrison, and seacoast guns; streamlined, smaller guns; strengthening and reinforcement of field and siege carriages; introduction of elevating screw on field carriages; permanent mortar beds; limbers with tongues for hitching horses abreast; caissons

American artillery—unstandardized equipment / cannon made in America / carriages and colors

Spanish artillery—French and English influences / carriages / fortress guns

Swedish artillery—ship cannon / identifying features

Advances in ammunition and firing technique—sabots and cartridges / firing at one stroke / hot shot / portfires / priming tubes / the gun crew

Spiking cannon

The art of gunnery

Reader's Guide to Illustrations

CHAPTER III THE NEW NATION, 1784-1835

CHAPTER IV THE APEX OF THE MUZZLE-LOADER, 1836-1865

Reader's Guide to Tables

CHAPTER I

The Beginnings

Artillery came early to America. Even before permanent colonies were planted, intrepid explorers dragged heavy pieces of ordnance through the wilderness as they sought to learn more of the new land and its possible sources of wealth. In 1539 Hernando De Soto landed in Tampa Bay to begin the epic march through the thick forests and swamps of the southern states which led to the discovery of the Mississippi River and the secret burial of his mortal remains beneath its waters. With him he took some bronze cannon of unknown size, but the effort of hauling these ponderous pieces of metal through the trackless forests proved too great a task even for such hardy souls as these. Powder was scarce also, and there was no opportunity to use these guns as De Soto had seen them employed in Mexico against the walled cities of the Aztecs. The Indians were impressed with them, however, and so with sudden inspiration he left them as a present to the important Indian chieftain Cosa, a gift which probably achieved the acme of impressive uselessness.

Farther west, Coronado dragged seven bronze guns with him as he marched north from Mexico in 1540. On the way America's first recorded artillery accident occurred. A gunner lost his hand when a cannon went off before he had completely withdrawn the rammer. Shortly thereafter Coronado abandoned four of the guns at Sia and pushed on with only three small pieces. These proved too light to batter down the Indian cliff dwellings he encountered and left him calling for a good old-fashioned catapult.

Despite these discouraging experiences, the Spanish exploring expeditions continued to take artillery with them. One of the last of the major expeditions, which penetrated and settled New Mexico under the leadership of Don Juan de Oñate in 1597 and 1598, was equipped with six pieces of heavy ordnance. These were three bronze culverins bearing the royal arms and dated 1585, two small bronze breechloading pieces and a little iron breechloader called an esmeril.

Artillery in the Colonies

BUT OÑATE was concerned with colonization as well as exploration, and it is with the establishment of permanent colonies that artillery really assumed an important place in America. When the French Huguenots built Fort Caroline near present-day Jacksonville, Florida in 1564, they brought with them bronze pieces bearing the arms of France. It was one of these guns that fired the first warlike cannon shot in American colonial rivalries when it opened fire on Menendez' Spanish fleet in 1565. The Spanish in turn proceeded to build their own forts in Florida, and they armed them with as many as ten guns each to fight off the English freebooters who ranged the Spanish Main preying upon the treasure fleets.

The English soon planted colonies, too, and they brought their own artillery. The Lost Colony of 1587 on Roanoke Island boasted falcons, sakers, and small breechloaders. The first fort at Jamestown mounted falcons and demi-culverins in 1607, and a year later the redoubtable Capt. John Smith with tongue in cheek generously offered an Indian four of the latter if he would carry them off. (They weighed about 3,400 pounds each.) The Pilgrims had only small guns of the minion and saker class when they founded Plymouth in 1620, but the wealthier colonists who settled Boston a few years later were much better equipped with cannon ranging up to one 18-pounder culverin.

The Dutch who settled New Netherlands were even better equipped, and Dutch artillery was among the finest made in the early 17th century. Nevertheless New Amsterdam's twenty guns, including 24-pounder "demi-cartoons," were not enough to prevent capture by the English and the removal of the Dutch influence on the history of American artillery.

The Dominance of English Artillery

With the capture of New Netherlands, control of land in what was to become the continental United States was divided among England, France, and Spain. This territory was never more than a sideshow for Spain, which was more interested in the Caribbean and Central and South America.

England and France were in deadly earnest, but it was England which held the most highly populated areas, and consequently English artillery dominated the 17th century as Spanish guns had held sway in the 16th.

From the English colonies and particularly from the Massachusetts Bay Colony at Boston stem some of America's earliest artillery traditions. Here was the cradle from which sprang much of the leadership of the Continental artillery in the Revolution and the early United States Army. Samuel Sharpe was the master gunner of Massachusetts Bay, appointed March 3, 1628 at a salary of ten pounds a year to oversee the planting of the ordnance and other matters pertaining to the artillery. Later came John Samford and Maj. Edward Gibbon. Then, in 1638, the Ancient and Honorable

English Cannon of 1646

Cannon	Caliber in inches	Weight of shot in pounds	Length of tube in feet	Weight of tube in pounds
Robinet	1¼	¾	3	120
Falconet	2	1¼	4	210
Falcon	2¾	2¼	6	700
Minion	3	4	8	1,500
Saker	3½	5¼	9½	2,500
Demi-culverin	4½	9	10	3,600
Culverin	5	15	11	4,000
Demi-cannon	6	27	12	6,000
Cannon	7	47	10	7,000
Cannon royal	8	63	8	8,000

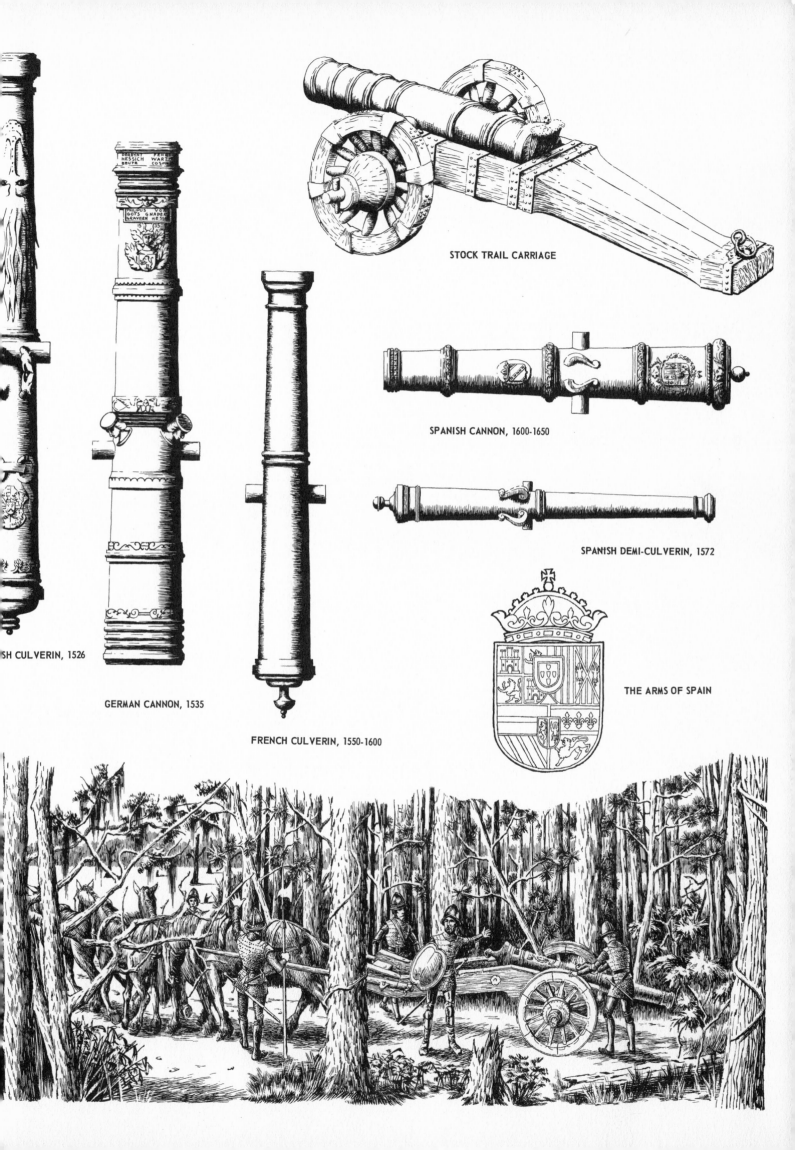

STOCK TRAIL CARRIAGE

SPANISH CANNON, 1600-1650

SPANISH DEMI-CULVERIN, 1572

SH CULVERIN, 1526

GERMAN CANNON, 1535

FRENCH CULVERIN, 1550-1600

THE ARMS OF SPAIN

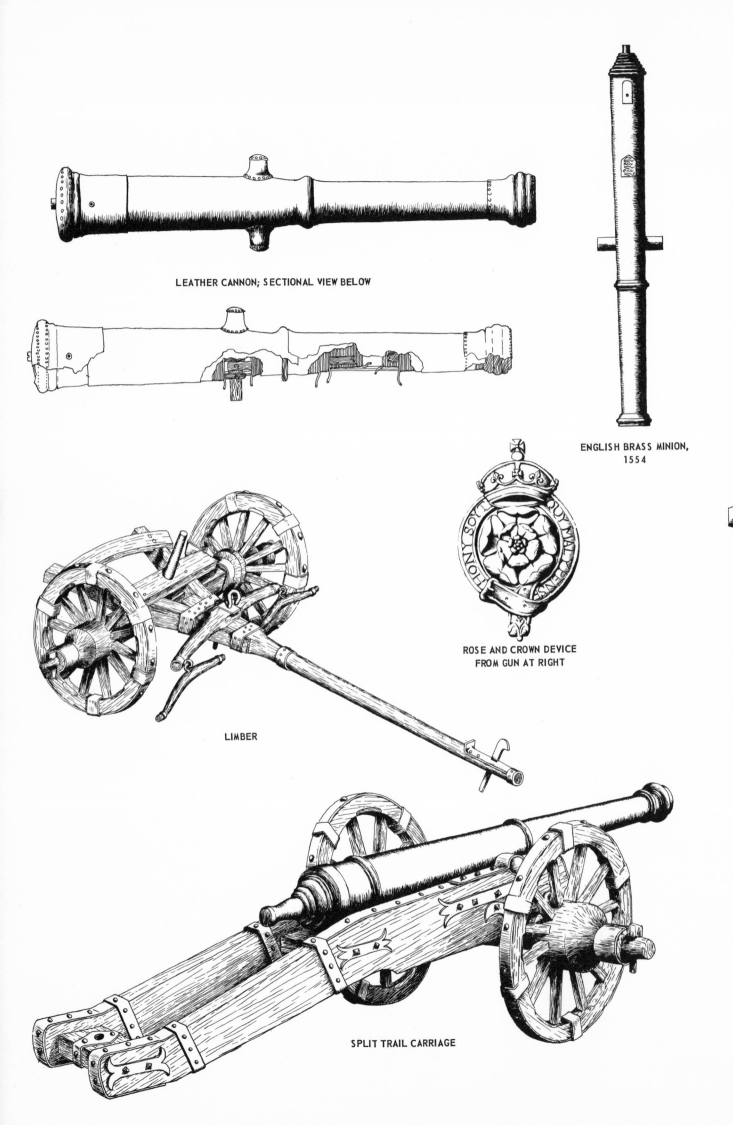

LEATHER CANNON; SECTIONAL VIEW BELOW

ENGLISH BRASS MINION,
1554

ROSE AND CROWN DEVICE
FROM GUN AT RIGHT

LIMBER

SPLIT TRAIL CARRIAGE

ENGLISH BR
DEMI-CANN
1542

Artillery Company of Boston was established by colonists from the original London company to propagate the art of shooting in the New World and to teach "scholars of great gunnes." Today, more than three hundred years later, it is still active.

The Limited Uses of Artillery

The artillery pieces brought to the New World by the early explorers and colonists were designed primarily for use in defending or attacking fortifications. Such major battles as might take place could be expected to center around an enemy stronghold. If an engagement did develop in the wilderness, the trees and undergrowth would probably be too dense to use artillery effectively.

Thus the military leaders selected their weapons accordingly.

It was the recollection of walled cities in Mexico and Peru that prompted the Spanish explorers to drag artillery through the wilderness. The wheeled limber to support the trail of the gun carriage had not yet come into use and so De Soto and Coronado probably hitched their teams to shafts which fastened directly on the carriage trail. Then, again, they may have used an even earlier system in which guns were dragged short distances by simply attaching a rope to the trail transom or else they were loaded on wagons for longer journeys. Since some of the pieces that would have been considered suitable for these expeditions weighed over 5,000 pounds, up to thirteen horses were needed to haul them over rough ground.

How Guns Were Classified

THE GUNS themselves were usually cast in bronze and given names to indicate their size and function. Most popular here were the culverin, demi-culverin, saker, minion, and falcon, but these terms covered a wide variety of guns. Generally they referred to the size of the bore. Most all nations used the same terms but applied them to slightly different pieces, and then the types could be modified still further by being described as long, short, fortified, light, extraordinary, bastard, and so on. And individual founders tended to follow their own ideas even after 1600. As a rule these guns were all long in relation to their calibers. There were also shorter battering pieces. These were properly called "cannon," and some of them are listed in the armaments of colonial forts also.

Thus, for example, in 1570 the Spanish fort of Santa Elena in Florida listed two cannon and one demi-culverin, plus "culverins of small bore," sakers, and falconets, and St. Augustine reported three reinforced cannon, three demi-culverins, two sakers, one demi-saker, and a falcon. Jamestown boasted twenty-four pieces of ordnance in 1609, mostly demi-culverins and falcons, mounted in the bastions and in the parade to command the gates. At Plymouth the Pilgrims had only two minions and two sakers along with two small breechloading pieces, to mount on the top deck of their fort overlooking the town and harbor.

As the years passed in Europe there were attempts to standardize sizes and patterns and reduce the number of different models. But here in America changes were slow, and pieces of the older pattern probably continued in use throughout the period.

Leather Guns

Early in the 17th century light guns designed for use in the field with troops were developed, and in Europe the practice of using these regimental or battalion guns in battles of encounter spread rapidly. Generally these were iron or bronze 4-pounders, but there was also an interesting experiment with so-called "leather guns." These pieces consisted of a wrought copper tube screwed into a brass breech and hooped with iron bands. This tube was successively covered with wrappings of mastic bound with layers of cord, coated with plaster, covered with leather boiled and shrunk to size and finally varnished. Only light charges could be used in such guns, and they could not be shot too often or too rapidly because the heat tended to destroy the wrappings.

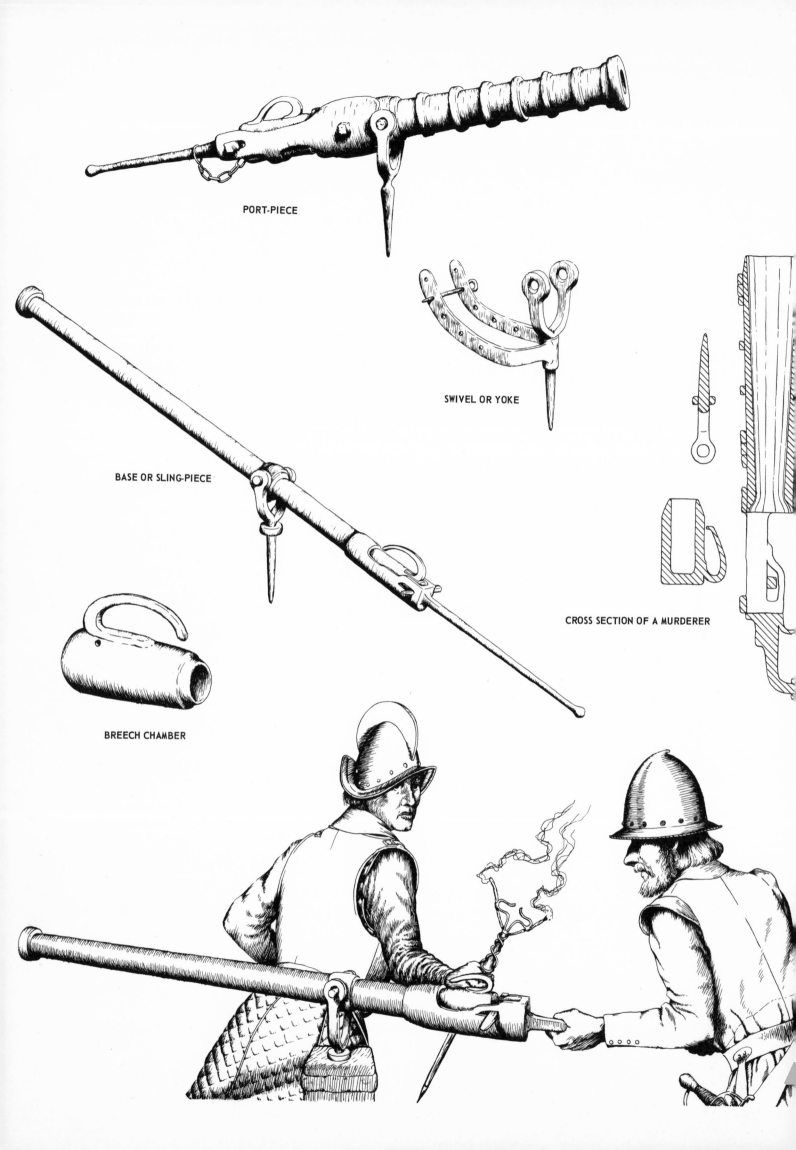

PORT-PIECE

SWIVEL OR YOKE

BASE OR SLING-PIECE

CROSS SECTION OF A MURDERER

BREECH CHAMBER

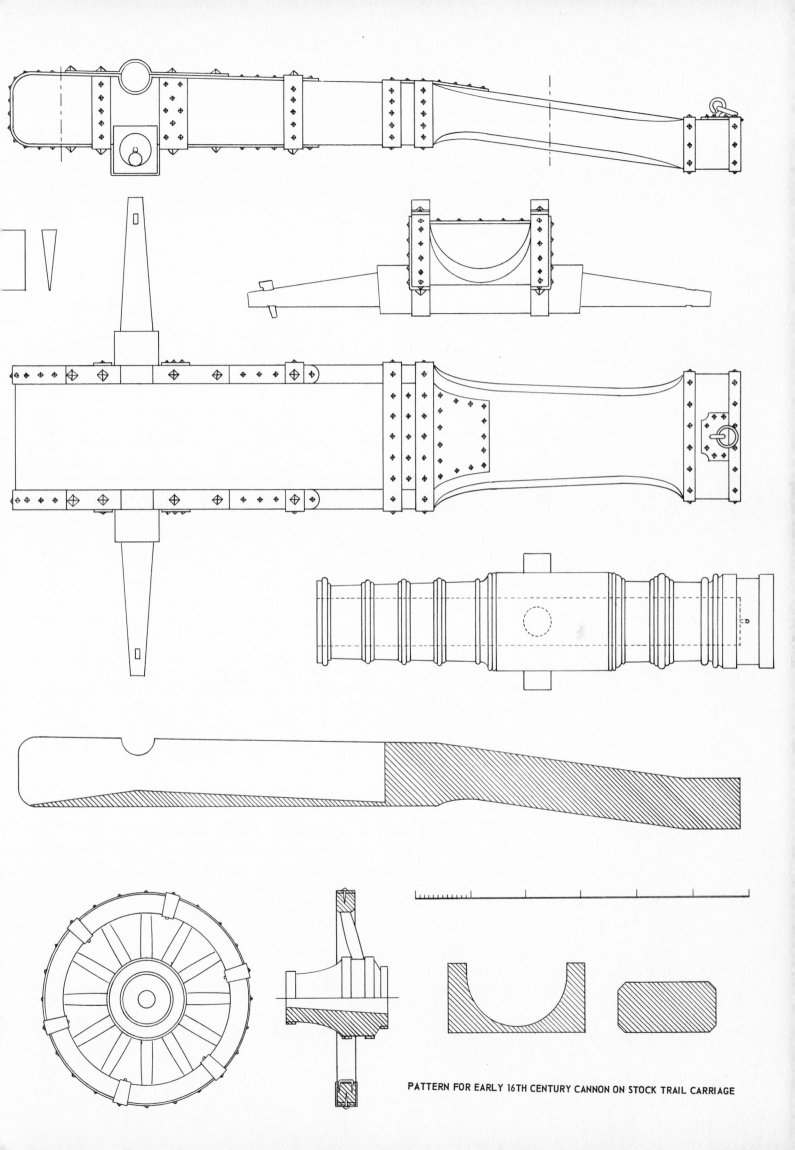

PATTERN FOR EARLY 16TH CENTURY CANNON ON STOCK TRAIL CARRIAGE

Ranges of Spanish Cannon of the Early 17th Century in Paces

Cannon	Point-blank range	Extreme range (45°)
2-pdr falcon	330	3,200
4-pdr falcon	400	4,000
6-pdr saker	450	4,500
8-pdr demi-culverin	500	5,000
10-pdr demi-culverin	550	5,500
12-pdr demi-culverin	600	5,700
15-pdr culverin	650	6,180
18-pdr culverin	700	6,700
20-pdr culverin	720	7,200
22-pdr culverin	800	7,355
12-pdr third-cannon	500	3,300
12-pdr demi-cannon	600	4,380
20-pdr demi-cannon	700	5,389
25-pdr demi-cannon	750	5,600
30-pdr cannon	800	4,900
35-pdr cannon	850	4,834
40-pdr cannon	900	4,792
45-pdr cannon	950	4,700
50-pdr cannon	1,000	4,660

Nevertheless they were an imaginative experiment in the search for mobility.

Always interested in new weapons, the American colonists followed this experiment closely. There are no records to indicate if any leather guns actually reached this country, but on November 11, 1647 the General Court of the Massachusetts Bay Colony passed the following resolution:

> For ye more easy & speedy transporting of great artillery . . . as also upon some suddaine designe to mount for advantage in an enemies workes, this co^rte doth ord^r, y^t, y^r be, by direction of ye maior gen^rall, 3 or 4 leath^r gunns, of sev^rall sizes, sent for to England by ye first opportunity, at ye charge of ye country, wch if found good & profitable, may give light & incouragm^t for ye procuring or making of more.

Swivel Guns

In addition to their large pieces the early colonists also had a series of small breechloading swivel guns. Such weapons had been used in Europe for mounting on the gunwales of ships and smaller vessels. Here they were popular for the defense of fortifications as well. Their small size made them easily transportable. They could be fired by a crew of two or three men, and since they were designed for rapid fire at close range, the crew did not need to be highly skilled in the more obscure facets of the gunner's art.

These swivel guns were ideal for defending a small fort against an assault by Indians, and they were used in large numbers throughout the colonial period. The "Lost Colony" on Roanoke Island boasted at least four or five of them in 1587. The Pilgrims at Plymouth mounted two on their fort and four more on a little platform in the center of the town, where they could command the streets and the gates of the stockade. Fourteen were listed in the vicinity of Jamestown in the census of 1624-1625, and still more were sent to Virginia in succeeding years. The same situation held true throughout the other colonies even into the eighteenth century, when the breechloading models were superseded by muzzle-loaders.

Many names were applied to these popular weapons. Among them were patarero, portingal base, base, drake, port-piece, stock-fowler, sling-piece, and murderer. Basically, however, there were two types. The bases, portingal bases, and sling-pieces were long guns with barrels about 30 times the diameter of the bore for the bases and 12 times for the sling-pieces. Normally the bore of these pieces averaged from 1½ to 2 inches. The second basic type were shorter guns often with larger bores. These comprised the murderers, port-pieces, fowlers, and drakes. Among this class the barrel averaged 8 times the diameter of the bore, which might be as large as 3 or 3½ inches. In some, the bore expanded from breech to muzzle in an effort to give a greater spread to the scatter shot they normally fired.

There were certain characteristics which all had in common. They were designed to fit in a yoke and swivel shaped somewhat like a modern oar-lock. This in turn was usually mounted in a stout post set in the ground and perhaps banded with an iron strap at the top for greater strength. The earliest types were most often made of wrought iron bars or a sheet of iron welded to form a tube and strengthened with hoops shrunk on as reinforcements. Later specimens were usually made of cast iron or brass. Each was equipped with a separate breech chamber shaped like a modern beer mug and containing a touchhole near its base. The mouth of this chamber was necked in slightly so that it could be inserted into the rear of the bore before it was locked in place with a key or wedge.

Loading one of these swivel guns was a simple and quick procedure. The shot was inserted into the rear of the bore followed by the chamber full of powder. One man could then hold the piece

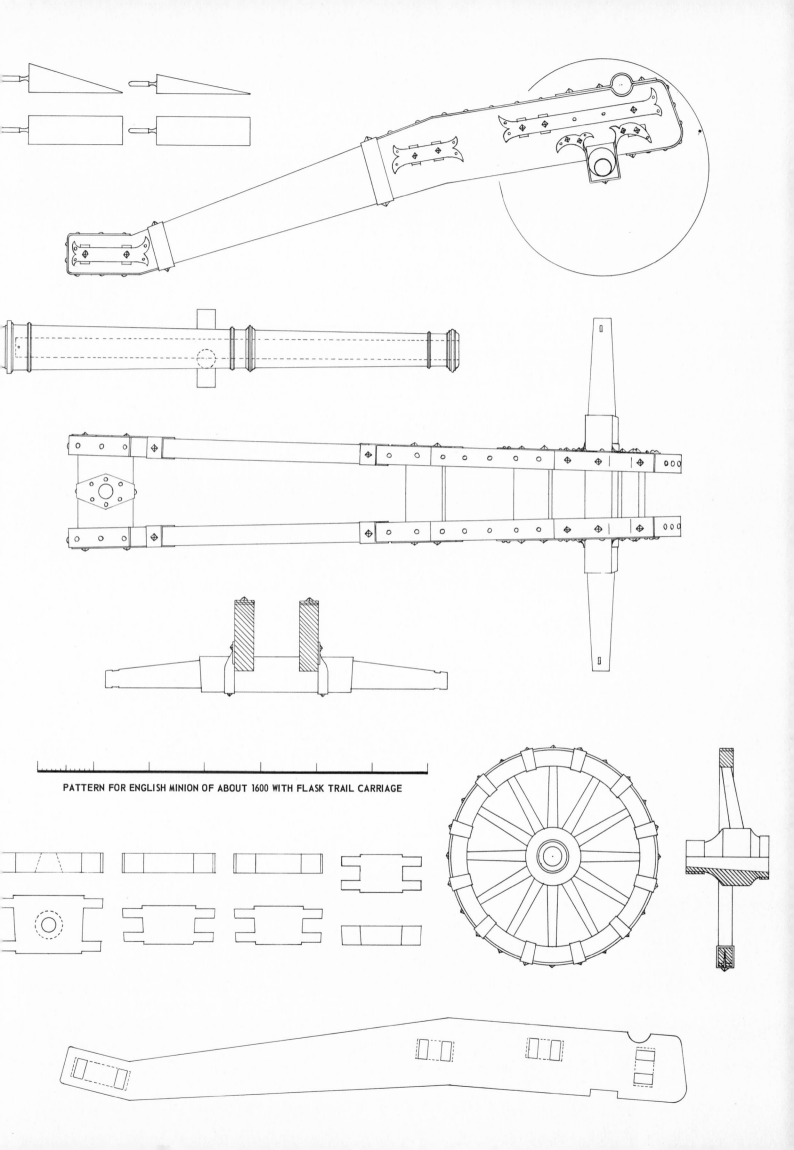

PATTERN FOR ENGLISH MINION OF ABOUT 1600 WITH FLASK TRAIL CARRIAGE

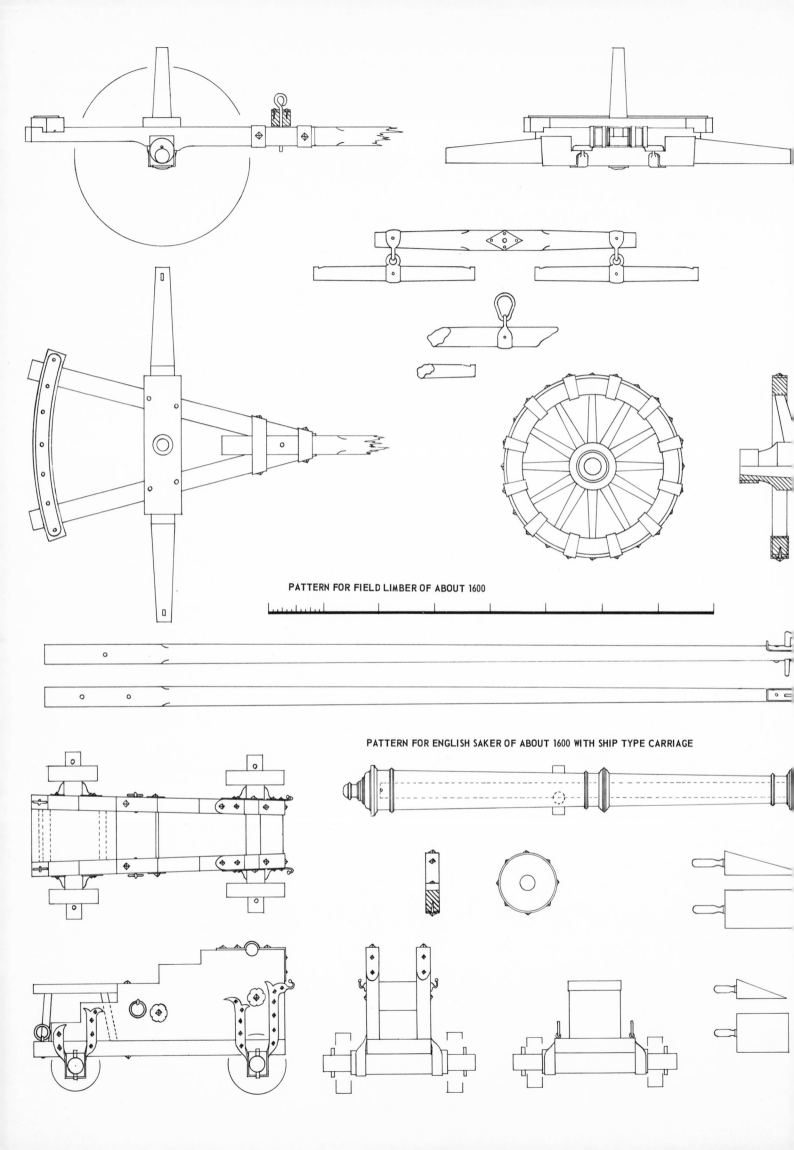

PATTERN FOR FIELD LIMBER OF ABOUT 1600

PATTERN FOR ENGLISH SAKER OF ABOUT 1600 WITH SHIP TYPE CARRIAGE

by the tail to train it while a second set it off. Extra chambers were often kept ready for greater speed in loading and firing. If available, a third man could be kept busy loading the empty chambers as the crew fired them. With this ease and speed in loading and firing plus their light weight and easy portability, it is small wonder that these little swivel guns remained so popular for American fortifications. Bigger guns were needed for sieges or for antiship or counterbattery work, but for antipersonnel use, which was the primary need of the frontier fort, these pieces were ideal.

Given the almost infinite variety of 16th and 17th century cannon, it is nearly impossible to offer valid generalizations concerning sizes, materials, or colors. Except for the wrought iron swivel guns and the leather cannon, almost all artillery of this era was cast in bronze. The iron pieces were painted black as a rule, though some appear to have been coated with red lead only. Bronze pieces were normally kept clean. They might be polished brightly in some garrison situations, and they might become quite dull in frontier posts or on an expedition. As a rule, however, they would not have been allowed to collect a green patina from oxidation.

The Evolution of Carriages

THE CARRIAGES on which cannon were mounted evolved slowly over the years. Among the guns used by De Soto and Coronado some or all may well have been mounted on carriages with a solid stock trail on which the breech of the tube rested. This was the earliest form of the wheeled carriage, and while it did provide more mobility than the older stationary beds, changes in elevation could only be made by raising or lowering the trail. By 1550 the latest guns in Europe all had carriages made of two principal members called "flasks" fastened together by cross pieces called "transoms." The breech of the tube now rested on one of these transoms and could be raised and lowered by means of a wedge, or "quoin." This greatly eased and speeded the aiming of the gun. The flask carriage was also much lighter than the solid stock and thus easier to move about. On most of the Spanish and French carriages of this type there were pivoted shafts attached near the trail transom which could be locked in position with a key for hitching to the team or folded back parallel to the flasks when the piece was in battery. Finally, sometime before 1600, the limber was invented and still greater efficiency achieved for the movement of the gun. The limber in its simplest form was a two-wheeled cart with shafts and a pintle mounted above the axle. When it was time to move a gun, the trail was lifted, the limber slid under it and attached by passing the pintle through a hole in the trail transom. The two together thus formed a four-wheeled wagon.

Fortress Carriages

Most of the guns in America, including those in the forts, were probably mounted on large two-wheeled carriages, which were far more maneuverable on rough ground or platforms and which could also be used in the field or for sieges. It is also probable, however, that some ship carriages were used, especially in forts such as the one at Plymouth, where the guns were taken directly from the *Mayflower*. These carriages with their solid sides and small trucks were acceptable for such permanent positions with walls for the attachment of training tackle and where the platforms were smooth. Without these conditions they were useless.

Favorite Woods

The carriages were made of wood, and the fasteners, reinforcing plates, tires, and other metal pieces were wrought iron. Oak was probably the favorite wood for most carriage constructions, but it was by no means universal. French artillerists of the late 16th century, for instance, preferred elm or walnut for the flasks; oak, elm, and walnut in that order for the transoms, and oak for the wheels. In America, cedar was also used upon occasion, at least by the French. Hickory also came into use on a limited scale, and in an emergency the best hardwood available would be pressed into service.

Carriage Colors

Both wooden and iron elements of a carriage were usually painted. In a few instances the wood might be left to weather naturally or perhaps treated with oil, but paint was used whenever possible for both protection and decoration. Colors tended to be bright and gay, and the irons were often of different hues. Red was widely popular for both wood and iron work. Blue and green also were used. Sometimes the carriages were painted in the colors of the ruling house or of the commander of an expedition. The Hapsburg rulers of the Holy Roman Empire, for instance, had many of their gun carriages painted black and red or black and yellow, the wood being black and the metal fittings red or yellow. Some Italian carriages were red with yellow mounts. In a few instances even more dashing effects were obtained by painting the wheels a different color from the rest of the carriage.

Carriage Dimensions

For all elements, both tubes and carriages, the caliber was used as a standard for measurement, much as it is with naval guns today. For a long piece such as a culverin, minion, or saker, the tube might be 30-34 calibers long. The flasks of the carriage for one of these pieces would usually be 24 calibers long or roughly about three-fourths of the length of the tube, and their width would taper from 4 calibers at the front to 2 at the trail. The transoms would be 1 caliber thick and 1 caliber wide. The front transom would be 3½ calibers long, the middle transom 4 calibers, and the

Data on Principal Continental European Cannon as Given by Ufano, 1621

Name	Caliber in pounds	Length in calibers	Point-blank range in paces (bore level)	Common range in paces ("metal level")	Maximum range in paces
Esmeril	1	37	158	315	1,873
Ribauldequin	1¼	36	206	411	2,459
Falcon	2½	36	279	568	3,318
Saker	5	34	350	700	4,139
Demi-culverin	10	33	450	900	5,373
Culverin	20	32	600	1,200	7,140
Double culverin (or Dragon)	40	31	582	1,364	8,167

trail transom 5. Each would be inset half a caliber into the flasks on either side. Wheels averaged half the length of the tube, though this rule was modified on some of the very long or very short guns. These proportions were recommended by Diego Ufano, whose *Artillerie* appeared in many editions and was widely used throughout Europe in the first quarter of the seventeenth century. Luis Collado had suggested generally the same dimensions in his *Platica Manual de Artilleria* of 1592 but added that when cannon were to be used in fortifications the wheels should be higher so that the axle would be equal to the height of the wall or the bottom of the embrasure over which the gun was to fire. In the field he felt that 9-caliber wheels were sufficient for a battering cannon of large bore. These were perhaps the most generally accepted proportions for carriage construction after the introduction of the flask or split trail pattern, but almost every founder, artillerist, or artificer felt free to introduce his own variations. Artillery was, after all, still a very inexact science, and there was plenty of room for experiments and imagination.

Ammunition in the Colonial Period

THROUGHOUT THE early colonial period the basic ammunition for almost all guns was a solid ball and a charge of black powder sufficient to propel it. The ball itself was usually cast iron, but in a few pieces round stones were still used as they had been in previous centuries. In the smaller guns the balls were occasionally made of lead. Each was designed to fit the bore loosely so that irregularities in the ball or the bore or fouling from the powder would not cause trouble. The difference between the diameter of the ball and that of the bore was known as "windage," and it was often a quarter of an inch and more.

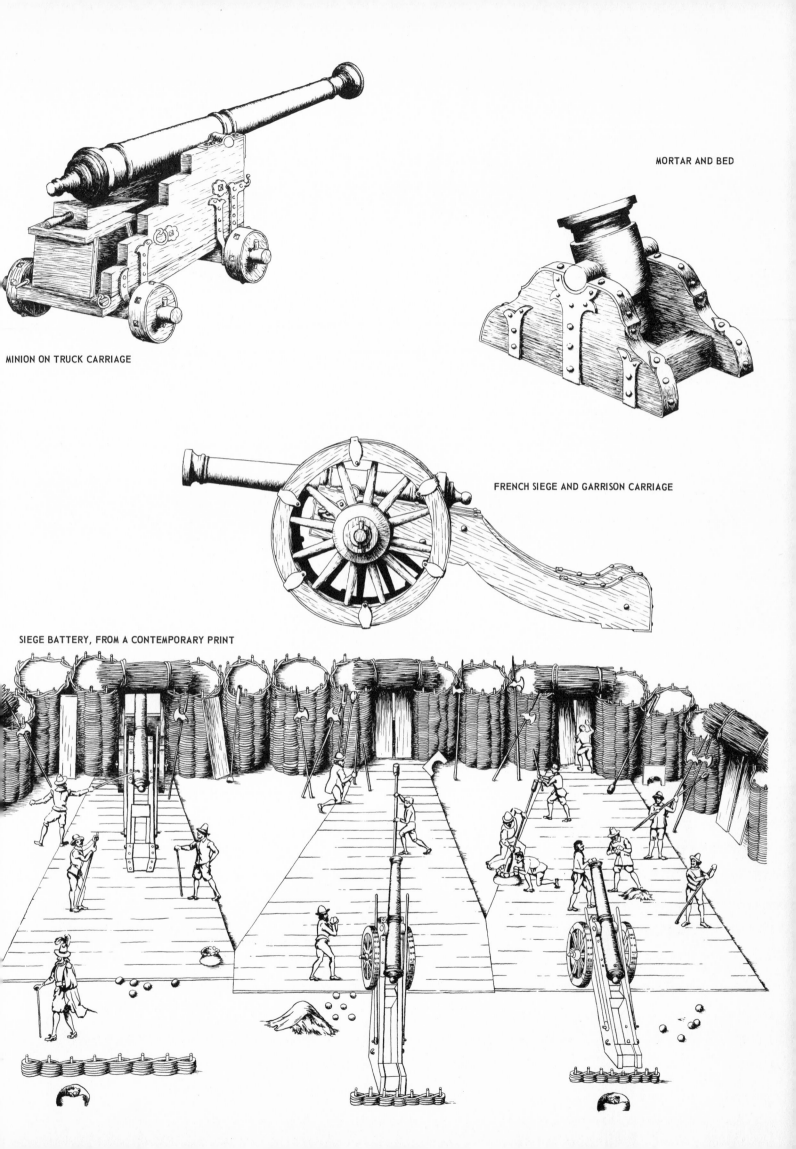

MINION ON TRUCK CARRIAGE

MORTAR AND BED

FRENCH SIEGE AND GARRISON CARRIAGE

SIEGE BATTERY, FROM A CONTEMPORARY PRINT

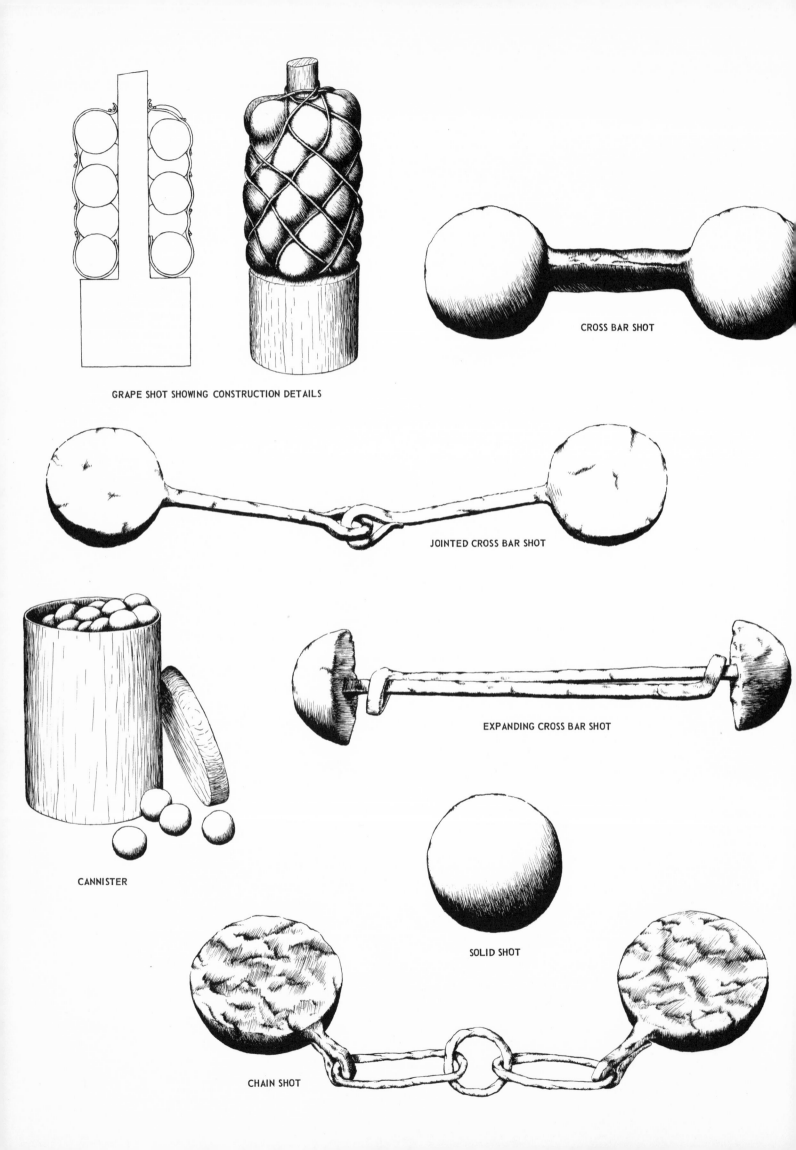

GRAPE SHOT SHOWING CONSTRUCTION DETAILS

CROSS BAR SHOT

JOINTED CROSS BAR SHOT

EXPANDING CROSS BAR SHOT

CANNISTER

SOLID SHOT

CHAIN SHOT

Gunpowder

Black powder was a mixture of saltpeter, sulphur, and charcoal. Proportions varied, but by 1650 the formula was generally 7 parts of saltpeter to 1¼ parts of sulphur and 1 part of charcoal by weight. At first powder had been simply a dust of these three ingredients, but by the time the first of the explorers set out for America a process had been developed whereby the mixture was moistened and squeezed through sieves to produce grains. Powder intended for cannon was made with large grains, for small arms with medium-sized grains, and for priming with small grains. By graining or "corning" powder in this way the rate of combustion and therefore the explosive force was increased, and the dangerous explosive dust which arose every time the older kind of powder was handled was eliminated.

Most artillerists preferred to have the powder brought into the battery loose in barrels and then carried to the front of the piece in a sack, valise, or box from which they would ladle the proper amount into the bore. Occasionally, however, a cartridge was used, but only if rapid fire was necessary. The first of these cartridges developed during the second half of the 1500's. They consisted of canvas or linen sacks each holding a charge of powder. In loading, these sacks were untied at the neck, rammed gently to the bottom of the bore, and punctured by a pick thrust through the touchhole to insure that the powder would be exposed for ignition. Sacks made of other materials also appeared, including paper, parchment, and wool. None burned completely, however. All left smoldering remnants that might set off the next charge prematurely and that had to be cleaned out every few shots because they fouled the bore. Early in the 17th century projectiles sometimes began to be attached to these powder bags, and in such instances a form of fixed ammunition had appeared on the scene.

Types of Shot

In addition to the standard solid balls there were a number of other kinds of projectiles, some eminently practical, some wildly imaginative as inventors tested their theories in the new field of ballistics. Some shot was designed to break apart and scatter. This category included langrage—any collection of old pieces of iron tied in a bag or put loosely into the bore, cannister—a sheet metal cylinder filled with small lead or iron balls, and grape shot—a group of iron balls clustered around a central wooden spindle on a wooden disc and held together by a canvas cover and lashings.

There was shot designed to turn on its axis as it flew through the air and thus do great damage to a ship's rigging or cut a wide swath through ranks of men. These included bar shot, cross bar shot, jointed shot, expanding shot, chain shot and the like. The most practical of these were the cross bar shot and the chain shot. Indeed, William Simmonds gives credit to a cross bar shot's whistling through the air and striking down the limb of a tree for breaking up an Indian attack that threatened the lives of a party building the first fort at Jamestown in 1607.

Shells and Incendiaries

Among the artillerist's most consistent efforts was the attempt to devise satisfactory explosive and incendiary projectiles. One of the first methods (described in 1561) involved filling a hollow brass or iron ball with powder and tieing it in a sack which was in turn filled with a priming compound. A hole was made in the sack opposite the hole in the ball and plugged with a stick. When ready to use, the stick was removed. The compound could then be ignited by hand and the projectile either thrown as a grenade or placed in a mortar and fired as a bomb. In the latter case the flame from the charge would ignite the compound which would then burn around the ball until it came to the hole. Then, perhaps, the whole thing would explode according to plan. Later, tubes of iron or wood filled with a priming compound were hammered into the fuze hole of the bomb. When these were put into the mortar with the fuze facing the charge so that it would ignite automatically, the force of the explosion frequently drove the fuze right into the bomb and exploded it before it left the piece. This practice was soon abandoned. The next step, which did not come into general use until almost 1650, was to place the bomb in the mortar with the fuze facing the muzzle, light first the bomb fuze, then apply the linstock to the vent and pray there would be no misfire.

An incendiary projectile known as a carcass also was widely popular. It consisted of an iron ball, sometimes filled with powder, wrapped with highly inflammable material and tied in a bag. It was

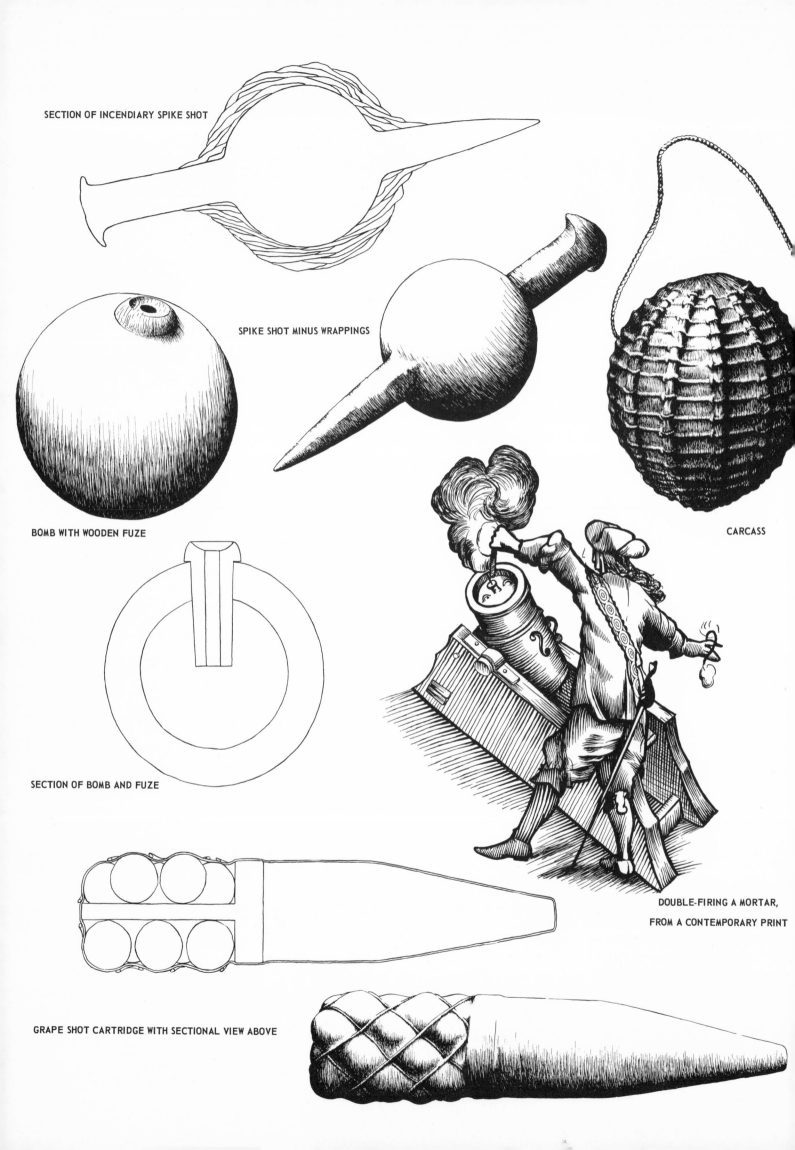

SECTION OF INCENDIARY SPIKE SHOT

SPIKE SHOT MINUS WRAPPINGS

BOMB WITH WOODEN FUZE

CARCASS

SECTION OF BOMB AND FUZE

DOUBLE-FIRING A MORTAR,
FROM A CONTEMPORARY PRINT

GRAPE SHOT CARTRIDGE WITH SECTIONAL VIEW ABOVE

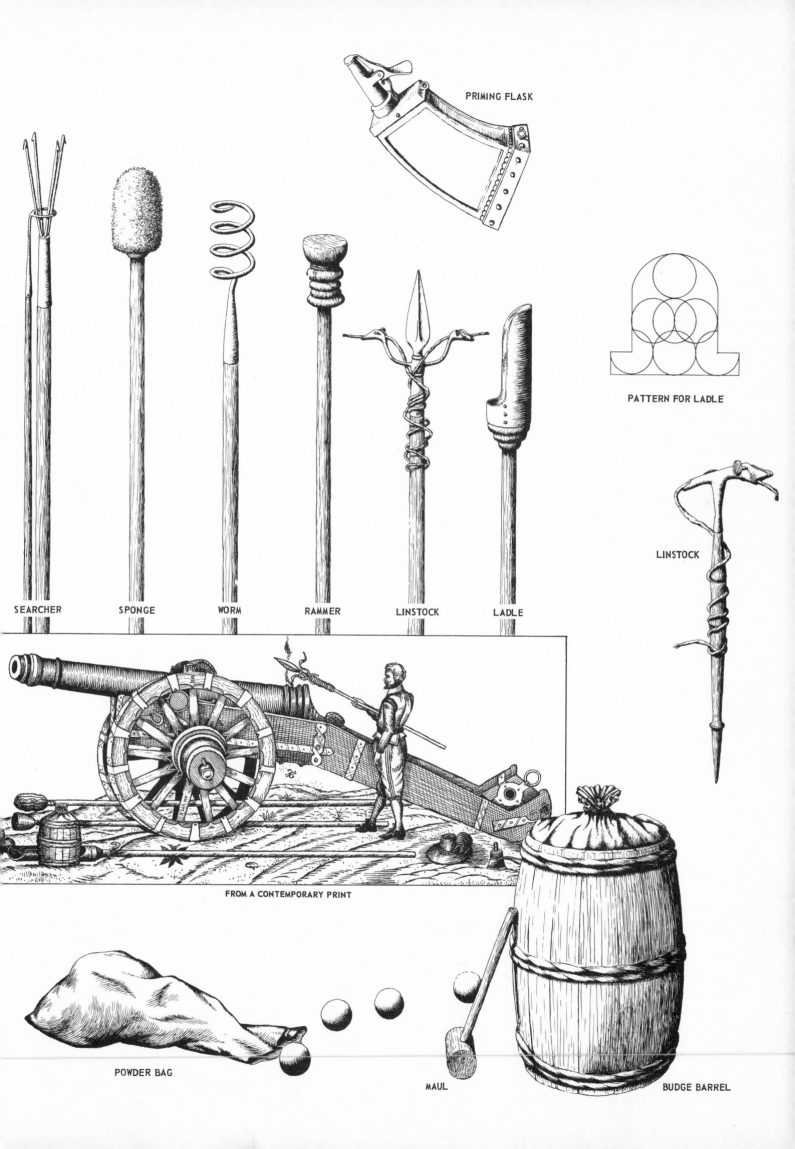

PRIMING FLASK

PATTERN FOR LADLE

LINSTOCK

SEARCHER   SPONGE   WORM   RAMMER   LINSTOCK   LADLE

FROM A CONTEMPORARY PRINT

POWDER BAG

MAUL

BUDGE BARREL

vastly effective against ships and wooden buildings. There was also spike shot, a ball wrapped with incendiary material and transfixed with a sharpened bar in the hope that it would stick fast in a wooden wall and set it afire. Imagination was the only limit to the variety of projectiles.

Duties of the Artillerist

LOADING, AIMING, and firing an early muzzle-loading cannon was a complicated process requiring the assistance of several men and the use of special equipment. The full gun crew as it developed in the 18th century and as it is known today had not yet evolved. For most of the period all but the largest guns were served by an artillerist or gunner and his "companion," who had the help of as many "pioneers" as they needed to shift the gun's position.

Loading and Firing

Upon approaching the gun, the artillerist first blew upon the wick or "match" held in the jaws of the linstock with which he would later fire the piece to make sure there was a glowing coal. Then he stuck the pointed end of the linstock in the ground near the breech of the gun, advanced to the muzzle, and picked up the sponge with its covered lambskin head. Using the sponge dry, he cleaned the bore, striking the muzzle sharply as he withdrew to shake off the dirt. Next, turning to his companion who was standing ready with either a budge barrel, valise, or sack of powder, he took the ladle with its copper head gauged to hold exactly the proper charge for the piece, filled it from the container, inserted it to the very bottom of the bore, withdrew a couple of inches and turned it over, spilling the gunpowder into the bore. On the way out he tilted the staff of the ladle to make sure that all the powder had been emptied and that none would fall on the ground. Meanwhile his companion had walked to the breech and placed his finger firmly over the vent. They did not understand the safety factors involved in this act, but did it only to prevent the powder being forced up out of the large touchhole when the artillerist rammed the charge home. This he now did, seizing the rammer, pushing all the powder to the breech and giving it two or three sharp blows to make sure it was well seated. Next a tight-fitting wad of straw, hay, tow, or similar material was driven well home on top of the powder. If time permitted, the dry sponge was again used to clean the bore above the wad to make sure there were no loose grains of powder left which might cause trouble. Then the ball, which the companion had carefully cleaned to remove any particles of dirt, was wrapped with a little bit of tow, placed in the bore, and gently pushed home, the artillerist taking care to stand to one side of the piece in case of an accidental discharge. Finally another wad was placed on top of the ball, and the piece was loaded.

It was then up to the artillerist to move the gun into battery and aim it. He did this with the aid of the pioneers and their levers or handspikes. That done, he primed the vent with powder from his flask or horn, took up his linstock, again blew the match into a glowing coal, touched it to the vent and fired his piece. Between firings, the bore was swabbed with the sponge dipped in vinegar and water to cool it and to put out sparks. When cartridges were used, the worm had to be employed frequently to remove unburned pieces of the wrappings.

Aiming

It was in aiming a cannon that the gunner really needed his "art." First he determined the line of sight along the top of the tube by using a level consisting of a frame and a plumb bob and marked the points on the breech and muzzle with beeswax and wood. He then had the choice of aiming directly over these marks—which would

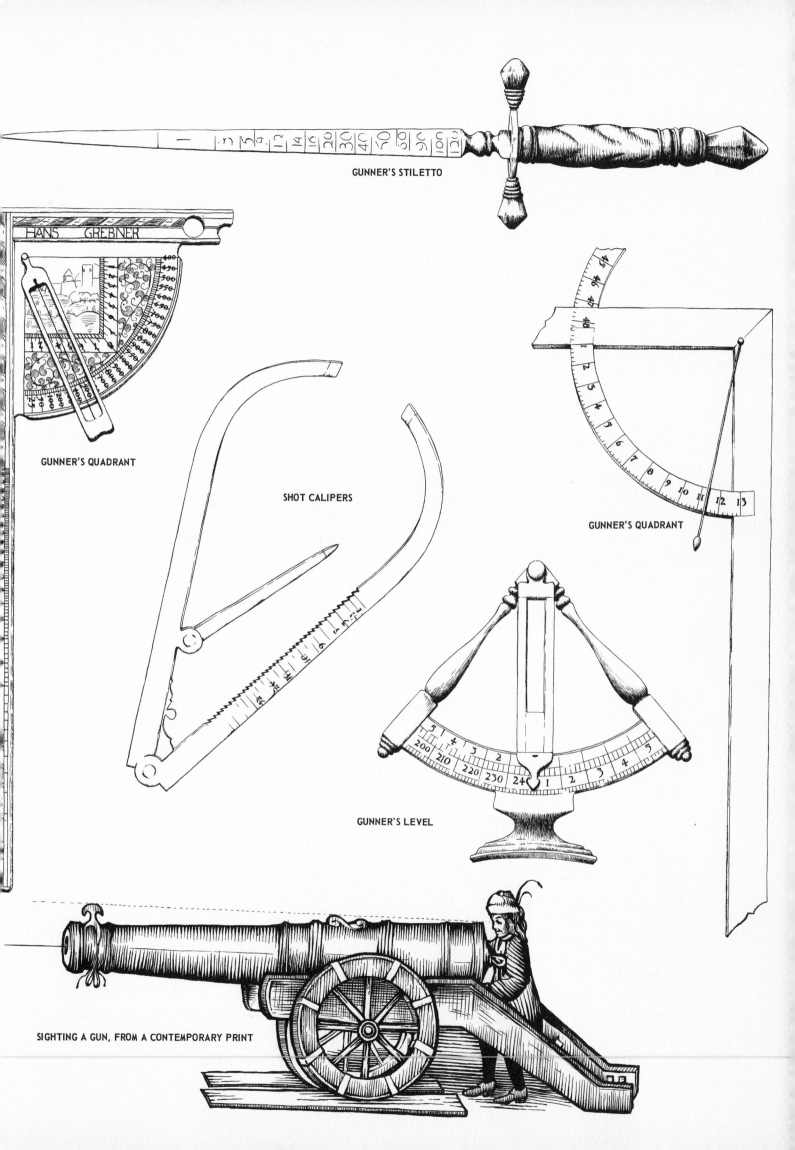

GUNNER'S STILETTO

HANS GREBNER

GUNNER'S QUADRANT

SHOT CALIPERS

GUNNER'S QUADRANT

GUNNER'S LEVEL

SIGHTING A GUN, FROM A CONTEMPORARY PRINT

give the piece a slight elevation because of the taper of the outside of the tube—or of raising the front sight sufficiently to compensate for the difference in radii so that he could aim in a line parallel with the bore.

Gunners preferred to operate at point-blank ranges. Elevating the muzzle of the piece to increase the range was known as "firing at random," and this it certainly was. The strength of powder might vary as much as 20 per cent and spoil the best of calculations. The large amount of windage allowed the ball to bounce along the inside of the bore and take off at an angle determined by the last bounce before it left the muzzle. Thus the greater the range, the greater the error. Nevertheless the gunners carried quadrants, the long arm of which they inserted in the muzzles of their pieces so they could read the angle of elevation as shown by the plumb line. This angle they checked against tables of ranges prepared according to the principles set forth by such scholars as Niccolo Tartaglia or Galileo but which failed to consider air resistance and so were never accurate. There were so many other chances for error, however, that the gunners never seemed to notice this. Anyway, if one had to shoot "at random" it was a good practical idea to have the first shot fall a trifle short and then keep elevating until the proper range was reached.

In addition to his quadrant and his level, the gunner often carried calipers for measuring shot to make sure he had the proper size for his bore. Instead of calipers he sometimes used a scale engraved on a rod or even on a stiletto blade for the same purpose. Really "scientific" gunners occasionally had linstocks with pierced heads from which they hung small geometrical tables with plumb bob and compass for surveying their field of fire.

Other pieces of equipment that belonged in a well-run battery included lead covers to be lashed over vents, buckets to hold the water and vinegar for sponging, priming wires to clear vents and open cartridges when they were used, searchers to check bores for holes or cracks, chocks for the wheels, and, most important, flint and steel for the gunner to strike a light if his match should go out.

CHAPTER II

The French Wars and the Revolution, 1689-1783

The years from 1689 to 1783 brought many changes in the history of artillery in America. All nations with colonies on this continent standardized their cannon and adopted systems of artillery, though some were more complete than others. Field artillery came into use, playing its role in battles of encounter, and, finally, true American artillery appeared with American artillerists taking their place as equals alongside their European counterparts.

Most important from the standpoint of the overall history of artillery was its standardization. By the beginning of the period all colonial nations had given over the old system of named pieces and had begun to designate them by size. They indicated the different guns by the weight of the solid shot that they threw. A gun throwing a 4-pound ball ceased to be a minion and became instead a 4-pounder. A culverin became an 18-pounder or a 20-pounder or whatever, depending upon its bore. It was a simple change, and yet its effect was sweeping. In one step it abolished all the confusion of the ancient days with guns of different bores bearing the same name. And it set the stage for further standardization as nations concentrated first on a specific number of bore sizes and then on similar proportions for all guns of the same caliber. Mortars followed a similar pattern, though they were designated by the diameter of their bores, since they threw bombs instead of solid shot, and bomb weights varied. Thus one had 8-inch mortars, 10-inch mortars, and so on.

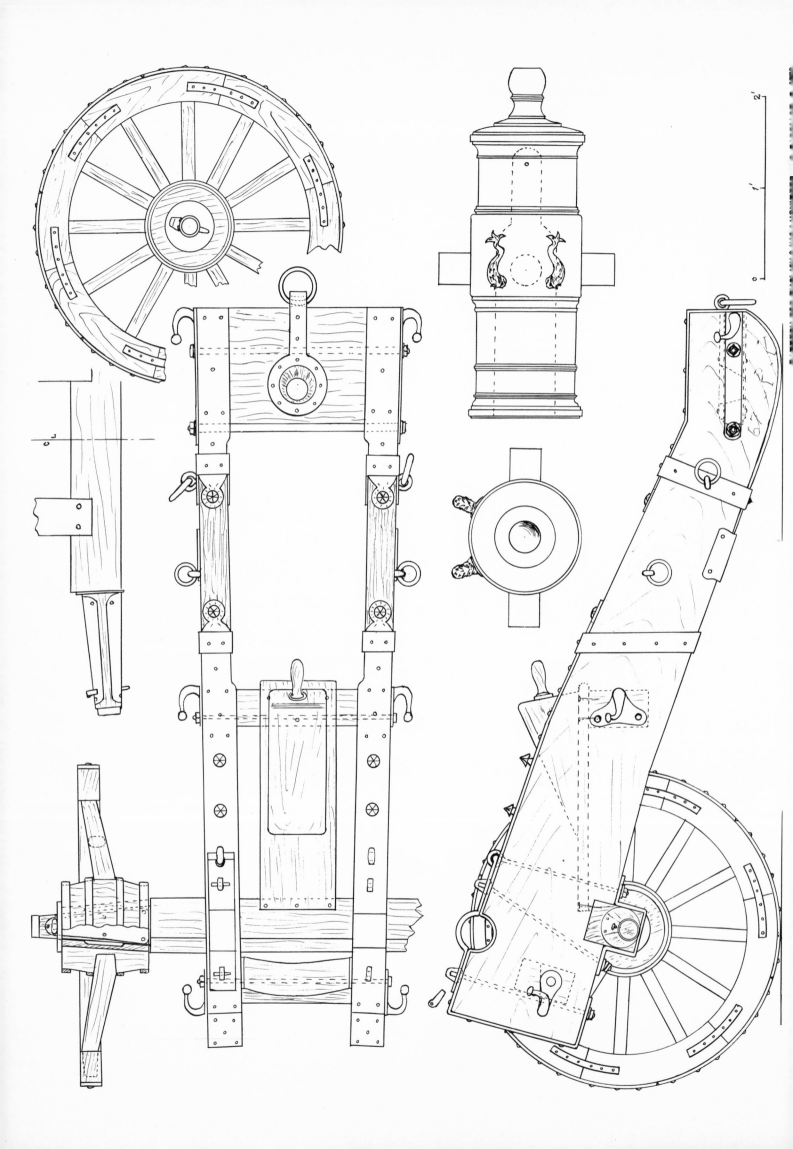

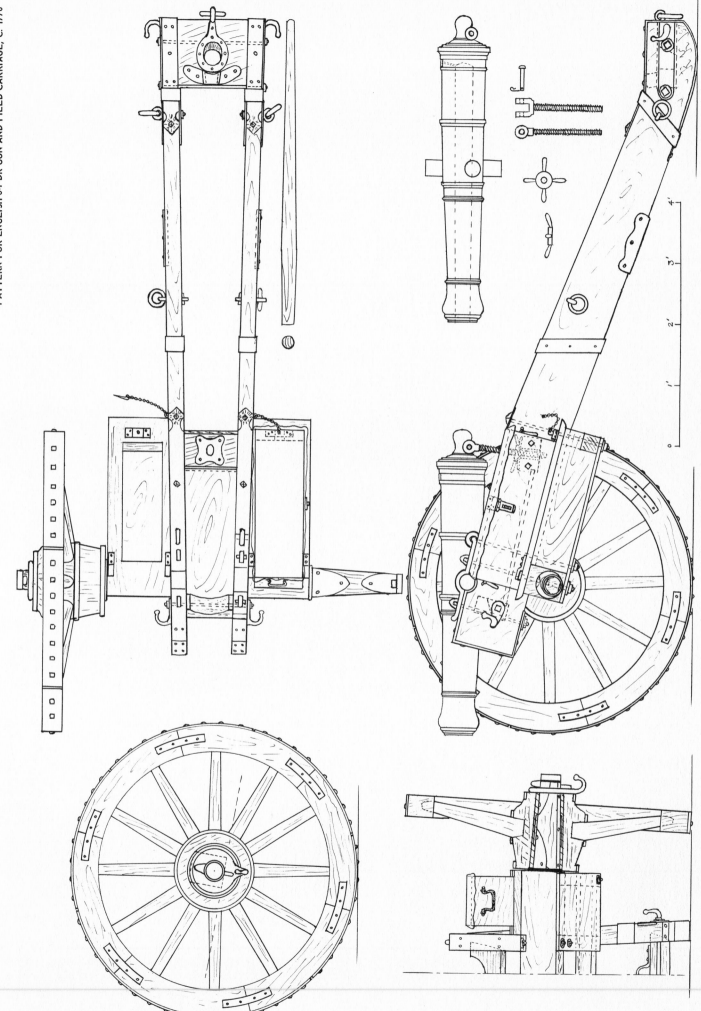

An intriguing new weapon, the howitzer, had also been invented in the Netherlands. This piece fired with a higher trajectory than the gun but lower than a mortar, and it was designed for two uses. Its primary mission was to throw shells behind an enemy's fortifications like the mortar, but it was more mobile than that chunky weapon. At the same time it could be used for langrage, grape, or cannister in antipersonnel work in the event of an enemy attack, as the mortar could not. Thus it commended itself to the attention of artillerymen all over Europe, and most nations moved to adopt the versatile arm as the period developed. Since it normally did not throw solid shot in this period of its use, the howitzer, like the mortar, was usually designated by the diameter of its bore.

The standardization of weapons that started with uniform calibers continued to complete systems of artillery. In 1732 General Vallière made a pioneering effort along these lines in France. General John Armstrong set similar standards in Great Britain in 1736, but these men went only part way. It remained for Jean Baptiste Gribeauval to systematize every phase of artillery materiel in France in 1767 and even to carry the work into artillery organization, creating probably the first complete artillery system adopted by any nation in the world.

Artillery in Siege Warfare

WHEN IT came to artillery employment, the new period started off much like the early colonial era. Artillery centered around forts. Hardy soldiers dragged a few small guns on campaigns, it is true, but this was rare. Colonials used cannon primarily to defend or attack fixed positions, and there was considerable activity in this field. Spain completed the Castillo de San Marcos at St. Augustine and improved its other forts in Florida. France built and armed Fort Carillon, Fort Duquesne, and other smaller posts and armed them with cannon. Great Britain built forts along the coast such as Castle William in Boston Harbor, Fort George, and others. In the interior they captured both Duquesne and Carillon, renamed them Fort Pitt and Fort Ticonderoga, and built many lesser posts as well. All of the major forts were armed with cannon, ranging in size up to the 18 32-pounders and 18 42-pounders at Castle William and down to the swivel guns in such tiny defenses as Washington's impromptu Fort Necessity of 1754.

Sieges there were, too. First, perhaps, was the Carolinian Moore's siege of St. Augustine in 1702. He had transported cannon for the purpose by sea, but his guns proved of no avail, and when his ships burned he had to abandon his artillery train to the defenders. General James Oglethorpe tried another siege of St. Augustine in 1740, but the balls from his cannon sank into the soft stone walls of the fort as if they had struck cheese instead of masonry, and he too came away defeated. Farther north, almost at the other end of the continent, British and American gunners successfully besieged the French fortress at Louisburg, the biggest fort in North America. And there were other sieges in which artillery played a lesser part.

Artillery in the Field

THEN CAME the American Revolution, and with it field artillery came into its own on this side of the Atlantic. There were sieges at Boston, Yorktown, and elsewhere, it is true, and in these sieges artillery played a major role. But there was little new in such use of

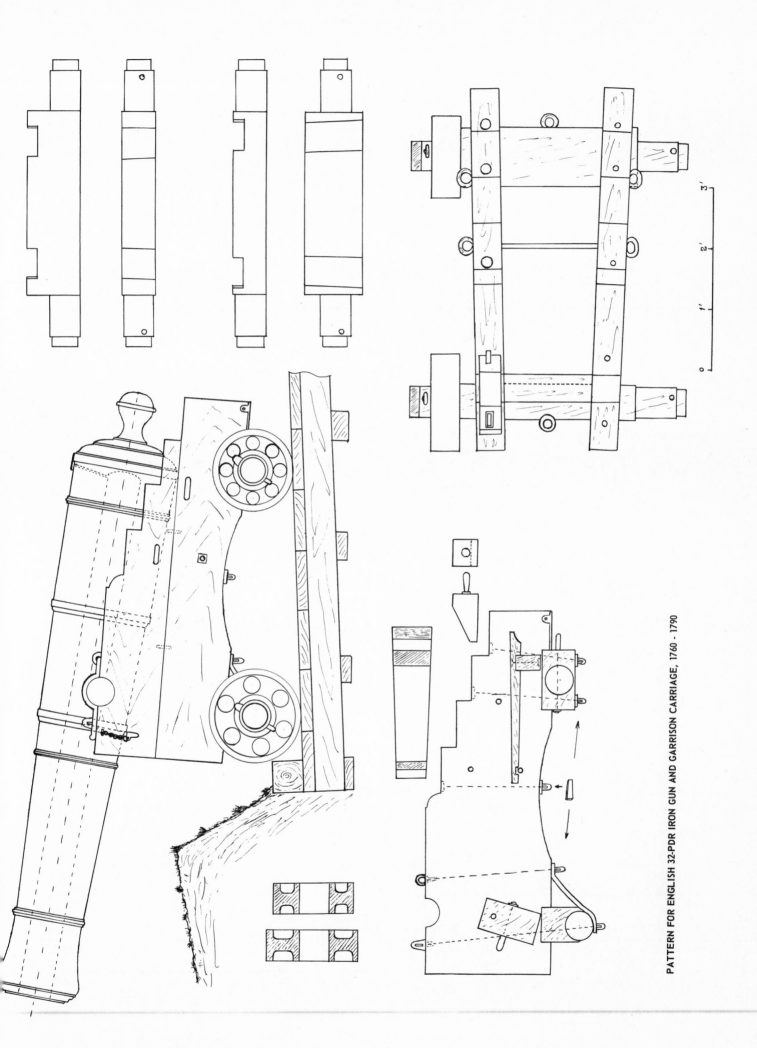

PATTERN FOR ENGLISH 32-PDR IRON GUN AND GARRISON CARRIAGE, 1760 - 1790

0   1'   2'   3'

cannon. It was battles of encounter in civilized areas with roads and cleared fields that brought artillery to the fore as an adjunct of infantry and often as a key to victory. This was true from the very beginning of the conflict. Two field guns accompanied the British forces that retreated from Lexington to Boston in the war's opening engagement, and gunners of the Royal Artillery Regiment together with some skilled German artillerymen firing English guns continued to serve Great Britain well throughout the war. On the other side mishandled American pieces contributed to

the defeat at Bunker Hill. It was a poor start for a nation's artillery, but under the administrative genius of Henry Knox the American gunners learned fast. Their confidence returned as they found they could meet the professionals on even terms. The culmination came at Monmouth in 1778. It was a drawn battle and a disappointment to Washington, but Knox was proud of his men, who had made the enemy admit that "no Artillery could be better served." With this accolade a new, wholly American artillery tradition had come of age.

British Artillery

FOREMOST AMONG all the types of artillery in America during the period of the French wars and the Revolution stood the British. Great Britain controlled the largest part of the land mass that was to become the United States, armed the most forts, and established the tradition which later American artillerists followed. At the beginning British cannon of all sorts were exceedingly heterogeneous in design and construction, and they remained diverse to a degree right to the end despite several attempts to standardize them. Both iron and brass were used for cannon throughout the period, and although the calibers of pieces were regulated early, lengths and weights remained more or less fluid. When General John Armstrong tested 24-pounders in the mid-1730's he found

six different lengths ranging from 8 feet to 10 feet 6 inches, and as late as 1764 the Board of Ordnance recognized three different lengths for brass or bronze 6-pounders and seven different lengths for iron 6's as currently standard.

Evolution of British Guns

Despite the variety of sizes, however, there are differences in design that enable students to trace the general evolution of English tubes in this period. The first distinct type that students recognize has been called the "rose and crown" pattern because of the raised design very like the 16th century Tudor device that is cast on the second reinforce of these guns. Some people, in fact, have been misled by this design to date the guns far earlier than they should have been because they thought them Tudor in origin. Actually they probably appeared some time in the second half of the 17th century and remained in use at least through the reign of Queen Anne, until 1714. All of the surviving specimens in America, in fact,

were made in Queen Anne's reign. These were long pieces in proportion to their bore with the second reinforce just a fraction longer than the first reinforce, and the chase twice as long as the second, giving them a very slender silhouette. The trunnions were centered on the low line of the bore, and the cascabel button, only slightly necked, had not yet developed the belt about its equator that was to characterize British tubes throughout the rest of the period. All known specimens of this pattern gun are iron.

Little is known about artillery design under George I, but in 1736 General Armstrong tried to bring some standardization to the manufacture of cannon by defining lengths for each size that he considered to be optimum after a series of tests. Supposedly they gave the greatest range and burned their powder charges in the most efficient manner. Generally these lengths ranged from 23 to 27 times the diameter of the bore, and so they were still long, slender guns.

John Muller, the great mathematician and artil-

ENGLISH GALLOPER CARRIAGE AND GUN

ENGLISH "ROSE AND CROWN" GUN, C. 1710

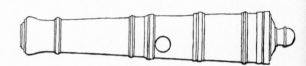

ENGLISH 6-PDR GUN, C. 1760 - 1770

ENGLISH 6-PDR GUN, C. 1740 - 1760

ENGLISH 6-PDR GUN ON FIELD CARRIAGE

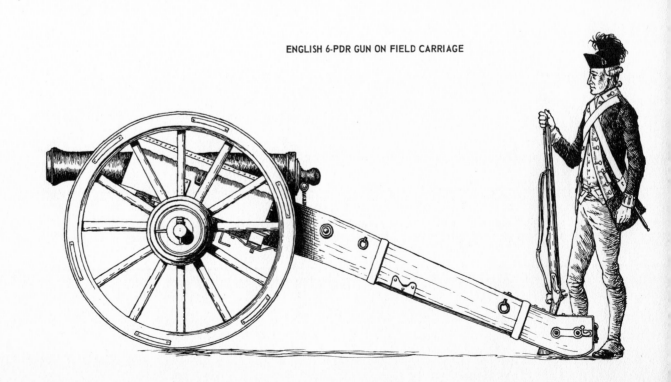

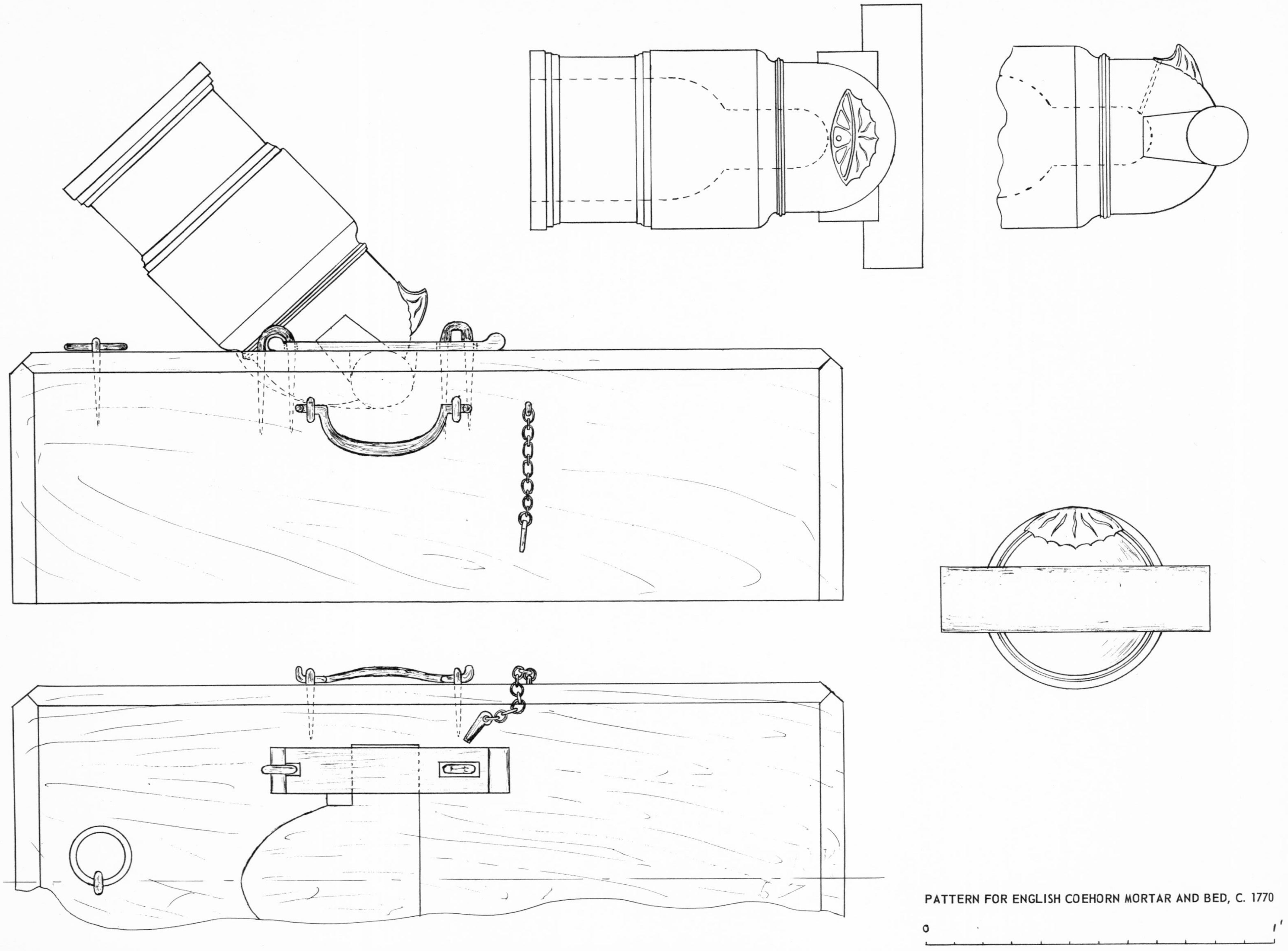

PATTERN FOR ENGLISH COEHORN MORTAR AND BED, C. 1770

Guns of the Armstrong System

Length in feet

Caliber	brass	iron
1½-pdr	6	—
3-pdr	7	4½
6-pdr	8	6½
9-pdr	—	7
12-pdr	9	8
18-pdr	9½	9
24-pdr	9½	9
32-pdr	10	9½

lery theorist at Woolwich, considered Armstrong's tests and designs the "least deficient" of all that had been made up to his own time. Nevertheless he felt that Armstrong's guns were much too long. So did many other artillerymen who had a chance to influence cannon design even before Muller. Within a very few years after Armstrong promulgated his designs, ordnance boards began to shorten guns, dropping them from more than 20 calibers long to more manageable lengths such as 15 to 16 calibers. Frequently they recommended guns of 15 calibers for field use and 16 calibers for ship service. Muller himself recommended guns of 14 and 15 calibers for these purposes respectively, but he was never able to get his ideas accepted by the Board of Ordnance. He also wanted to move the trunnions up so that they centered on the bore, and he advanced good scientific reasons for the new position. But this idea, too, was rejected, and British guns retained their trunnions on the low line of the bore well into the next century.

By the 1740's British cannon had assumed the general proportions they were to retain throughout the period. The belted cascabel button had appeared, and the mouldings had become generally standardized. Both iron and brass guns were in use. Nevertheless students can generally assign reasonably close dates to a given piece by looking at minor details. The breeches on earlier guns are apt to be more domed, for instance. In later years the breech flattens somewhat. Trunnions, which were one caliber long and one caliber in diameter throughout the period, taper slightly until about 1760, when they become straight-sided. Rimbases appear on trunnions in the 1770's, and on brass guns lugs for elevating screws first appear sometime late in the 1750's or early in the 1760's.

In all, the Board of Ordnance in 1764 recognized 11 different calibers of guns. There were ½-, 1½-, 3-, 4-, 6-, 9-, 12-, 18-, 24-, 32-, and 42-pounders. All but the 1½-pounder were made in iron. That small gun was cast only in brass, and there were brass 3-, 6-, 12-, 24-, and 42-pounders as well. There seemed to be sufficient variety for the most particular artillery officer to choose just the right size for any given job.

Mortars, Howitzers, and Petards

In addition to the guns there were also mortars and howitzers. Among the mortars there were 4½-inch or coehorn mortars named after their inventor, Baron Menno van Coehorn, a Dutchman. They were light pieces designed to be carried by two men and so were the most mobile of the regular mortars. Next came the 5.8-inch or royal mortars, and then the 8-, 10-, and 13-inch models, which were used primarily for siege work. In addition there were 2¼- and 3½-inch mortars apparently designed for throwing hand grenades. These light weapons do not appear on any official ordnance list that has yet been found, but there

Dimensions of All Sorts of Cannon, as Established by the Board of Ordnance in 1764

Nature		Pders.	Length	Weight	Caliber of the gun	Diam. of the shot
			F. in.	C. q. lb.	in.	in.
Brass guns	Heavy	42	9 6	61 0 0	7.3	6.68
		24	9 6	52 0 0	5.83	5.54
		12	9 0	29 0 0	4.63	4.40
		9	9 0	26 0 0	4.21	4.0
		6	8 0	19 0 0	3.66	3.48
		3	7 0	11 2 0	2.91	2.77
		1½	6 0	5 2 0	2.31	2.20
	Medium	24	8 0	40 1 21	5.83	5.54
		12	6 6	21 0 14	4.63	4.40
		6	5 0	10 1 0	3.66	3.48
	Light	24	5 6	16 1 12	5.83	5.54
		12	5 0	8 3 18	4.63	4.40
		6	4 6	4 3 14	3.66	3.48
		3	3 6	2 3 4	2.91	2.77

Nature	Pders.	Length		Weight			Caliber of the gun	Diam. of the shot
		F.	in.	C.	q.	lb.	in.	in.
	42	9	6	65	0	0	7.3	6.68
	32	9	6	55	0	0	6.42	6.10
	24	9	6	49	0	0	5.83	5.54
	24	9	6	47	2	0	5.83	5.54
	18	9	0	40	0	0	5.29	5.3
	12	9	0	32	2	0	4.63	4.40
	12	8	6	31	2	0	4.63	4.40
	12	7	6	29	1	0	4.63	4.40
	9	9	0	29	0	0	4.21	4.0
	9	8	6	27	2	0	4.21	4.0
Iron guns	9	8	0	26	2	0	4.21	4.0
	9	7	6	24	2	0	4.21	4.0
	9	7	0	23	0	0	4.21	4.0
	6	9	0	24	0	0	3.66	3.48
	6	8	6	23	0	0	3.66	3.48
	6	8	0	22	0	0	3.66	3.48
	6	7	6	20	2	0	3.66	3.48
	6	7	0	19	0	0	3.66	3.48
	6	6	6	18	0	0	3.66	3.48
	6	6	0	16	2	0	3.66	3.48
	4	6	0	12	1	0	3.21	3.4
	4	5	6	11	1	0	3.21	3.4
	3	4	6	7	1	0	2.91	2.77
	½	3	0	1	1	25	1.58	1.52

are surviving specimens which tests show to have been potentially effective weapons and not just toys or models. They may well have been designed as experimental infantry weapons like the modern trench mortars. If so, it seems that they would have performed well in this role, for they throw grenades accurately far greater distances than they could be hurled by hand. The howitzers came in 10-inch and 8-inch sizes. Both mortars and howitzers were chambered pieces; that is, there was a chamber for the powder charge that was smaller in diameter than the bore, and both mortars and howitzers were normally made of brass.

There was also one additional form of artillery called a petard. This was a simple conelike tube designed to be hung against the door of an enemy fortification so that it would blow the portal open when fired. Petards were dangerous weapons to use. They required at least one and sometimes several brave men to advance under fire, hang the piece in place, and then fire it. Occasionally petard attacks were successful, but generally they were expensive in lives lost and so were little used. A few are mentioned in colonial records, however, and so it is definite that some were present among British artillery stores in America.

Markings

Most British artillery tubes were marked. This is true, at least, of guns, howitzers, and mortars, and it may well have been true of petards as well. Some of these marks were largely decorative, but all tell a story when the student has learned to read them. First of all is the royal cypher, which often appears on the first or second reinforce and which indicates the monarch at the time the piece was cast. This appears on many brass pieces and on some iron guns as well. Through most of the period it consisted of the letters "GR" (for Georgius Rex) under a crown, but sometimes the numeral 2 or 3 was added to indicate which George was intended. The chase of brass guns also sometimes bore the badge of the master general of the ordnance. Both the royal cypher and the badge were normally cast in relief. A broad arrow indicating government ownership, however, was usually chiseled on the top of the tube as well. On brass pieces the name of the founder, the place of manufacture, and sometimes the date were often placed on the breech ring, and on all pieces the actual weight was usually chiseled either on the breech ring, the breech itself just above the cascabel button, or on the first reinforce ahead of the vent. This figure was inscribed in a series of three numbers indicating hundredweights separated by a dash or a dot as, for instance, 17-3-1. In this series the first number represents the full hundredweights, the second the quarters, and the third the remaining pounds. Since the British hundredweight actually weighed 112 pounds, the weight of the piece would be translated as follows: 17 × 112 plus 3 × 28 plus 1, or 1,989 pounds. A few iron guns that did not customarily bear the name of the founder on the breech ring did carry his initials on the face of one trunnion, but this

ENGLISH LIMBER

MONOGRAM OF GEORGE II

ARTILLERYMAN FIRING ENGLISH COEHORN MORTAR, C. 1770

ENGLISH HOWITZER ON CARRIAGE, C. 1770 -1790

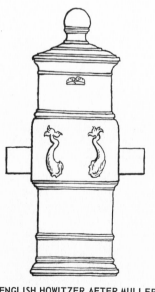

ENGLISH HOWITZER AFTER MULLER

SWIVEL GUN WITH SOCKET FOR TILLER, POSSIBLY 1770

ENGLISH SWIVEL GUN WITH MONKEY TAIL, C. 1750

ENGLISH 10-INCH MORTAR AND BED, C. 1760

ENGLISH 18-PDR SIEGE GUN ON CARRIAGE, C. 1770

was unusual. The makers of most iron cannon remained anonymous.

Brass vs. Bronze

Actually the terms brass and bronze as used during the 18th century and the opening years of the 1800's deserve some comment. In almost all the contemporary references the term used is brass. Bronze is almost never mentioned. Yet the alloy itself sometimes consisted only of copper and tin, which would make it bronze according to a modern definition. Sometimes zinc was added to the copper and tin mixture, and sometimes the alloy used was actually copper, brass, and tin. John Muller was perhaps the wisest of all when he avoided the issue entirely and referred to "gun metal" instead of either brass or bronze. Since it is impossible to know exactly the alloy used in any given instance, the terms brass and bronze will be used here as practically synonymous.

Carriages, Beds, and Mounts

By the 18th century the flask or split trail carriage had reached its full development for guns and howitzers but there were variations. The English version, for instance, had side boxes for ammunition on those intended for field guns. Some field guns and howitzers also had elevating screws for aiming while others still retained the old-fashioned wedge or quoin. Siege artillery had heavier carriages and lacked the side boxes. Howitzer carriages were a new development of this period. They generally resembled the carriages of the guns, but the trails were a bit thicker and shorter to allow for the greater elevation at which howitzers were customarily fired.

There was also another type of field carriage called a galloper. This was used for the lightest of the field guns, the 1½- and sometimes the 3-pounders. It consisted of a cart with two shafts so that a single horse could be hitched directly to it, and offering side boxes and cheeks for the actual mounting of the piece. Guns on galloper carriages were normally attached directly to infantry units. Colloquially the galloper carriage was sometimes called a grasshopper carriage to distinguish it from a light gun with a limber, which was called a butterfly. These were slang terms, however, and sometimes they were used very loosely.

Throughout the period British field carriages and rolling stock were painted a lead gray. The irons, including iron gun tubes, were painted black.

Specific fortress carriages also developed in this period. Well before it began the wheeled carriage had ceased to be used for guns in permanent fortifications, and ship type carriages had become standard, though there was still difficulty in traversing them rapidly. In the British service carriages for forts differed from ships' carriages by having small cast iron wheels or trucks instead of wooden ones.

Mortars were normally mounted on beds without wheels. The coehorn bed was the simplest of these, for it was just one great block of wood hollowed out to receive the breech of the mortar with straps of irons bolted over the trunnions. Normally it had handles for carrying. Bigger mortars still boasted solid beds, but these were built up from several pieces of wood, and they lacked the carrying handles.

Finally, there was one group of artillery pieces that needed no carriages. These were the swivel guns—the ½-pounders. Like their predecessors these pieces were mounted on a yoke swivel, but there was a decided difference. The swivels of the 16th and early 17th century had had a training tail manufactured as a part of the piece. This tail disappeared in the 18th century. Instead most swivel guns were cast with a typical cascabel button and used a separate handle called a monkey tail for the aimer to hold in firing them. Other swivels boasted a socket for the insertion of a wooden tiller for the purpose. Both monkey tails and pole tillers are known today only from documents and pictures. No contemporary specimens are known to exist.

Rolling Stock

In the field the British artillery was served by a well-organized system of rolling stock. There were wagons for carrying ammunition, powder carts for extra powder, two-wheeled tumbrels for general purposes, and travelling forges for field repairs. There were, of course, also limbers. These were little developed over the 17th century versions. They still consisted of a pair of wheels with a pintle on the axle for hooking through the trail transom of the cannon carriage plus shafts for hitching the horses. For British field guns there were a pair of these shafts so that the horses had to be hitched in tandem with one ahead of the other. For heavier guns tongues were used so that the horses could be hitched abreast.

These were the guns and carriages that fought

ENGLISH AMMUNITION WAGON

ENGLISH TRAVELLING FORGE

ENGLISH POWDER CART

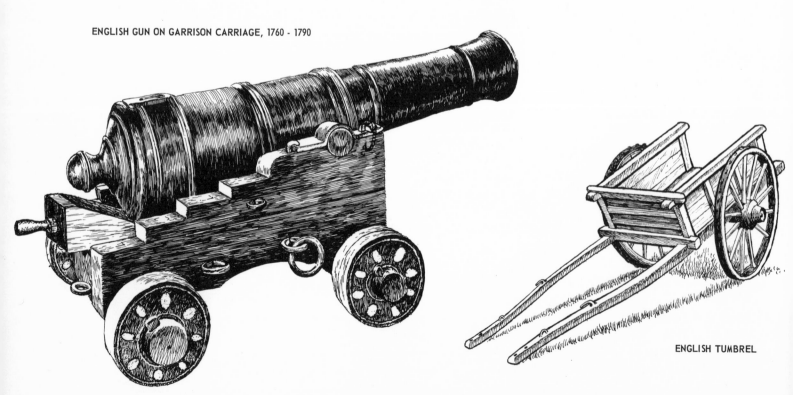

ENGLISH GUN ON GARRISON CARRIAGE, 1760 - 1790

ENGLISH TUMBREL

PATTERN FOR LIMBER FOR ENGLISH 6-PDR FIELD GUN, 1760 - 1790

throughout the French wars and the Revolution and continued as an inspiration to the newly formed American artillery. They were not the handsomest of the guns in the period; nor were they the most modern. Both of those distinctions fell to the French in different stages. But they were solid, well-designed engines that ably performed the duties assigned to them.

French Artillery

SECOND IN influence on American artillery history came the French. There were fewer French cannon in the fortresses that stood in what is now the United States, but American artillerymen received French guns during the Revolution, and in the next century the United States adopted the great French Gribeauval artillery system that first appeared during this period.

At the beginning of the period French artillery was completely heterogeneous. There were no standards whatsoever, and each founder or artillery official continued to cast cannon according to whatever pattern he felt best. Most guns intended for field use were brass. Fortress guns might be either brass or iron. Generally all were long pieces in proportion to their bore, and the brass ones were beautifully decorated with sculptured handles called dolphins for use in mounting and dismounting the piece, foliate cascabel buttons, a foliate band near the muzzle, often a scroll bearing the name of the gun itself, and various arms, frequently including the royal arms on the first reinforce.

The Vallière System

The confusion of designs and calibers was such that orderly artillerists longed for an end to it. When one of their number, General Vallière, was appointed by the King to take charge of all French artillery he quickly began to bring order to the jumble of ordnance materiel. In 1732 he standardized French gun calibers as 4-, 8-, 12-, 16-, and 24-pounders and added 8- and 12-inch mortars plus a 16-inch stone-throwing mortar. He specified lengths, proportions, and weights for each and even decreed the methods of manufacture. Going still further, he described the decorative details that should appear on the barrel. All guns of the same caliber were to be alike regardless of function. There was no distinction between field, siege, and garrison service.

The result was a series of handsome guns as well as standardized models. All Vallière guns were brass, and they were long and slender. They ranged from 20 calibers long for a 24-pounder to 25 calibers for a 4-pounder, and for figuring proportions, this length was broken into seven equal parts. The first reinforce took two of these parts, the second reinforce one part, and the chase the remaining four. The trunnions were located at the front end of the second reinforce, and what is more, they were to be one caliber long and one caliber in diameter and centered on the low line of the bore.

The decoration was to be functional as well as ornamental. Most notable for their dual roles were the breech faces and cascabel buttons. Each caliber usually had a different design so that a gunner could recognize the size of piece simply by a glance at the breech. The 4-pounder had a face in a sunburst; the 8-pounder, a monkey's head; the 12, a rooster head; the 16, a Medusa head; and the 24, Bacchus. The dolphins were actually sculptured in the form of dolphins, and there was a series of decorative and informative low relief sculpture. From the muzzle to the breech these latter designs

Vallière System Guns

Caliber	Length (in calibers)	Weight (in livres*)
4-pdr	25	1,150
8-pdr	24	2,100
12-pdr	23	3,200
16-pdr	22	4,200
24-pdr	20	5,400

* A livre weighs 1.1 lb. in English or American weight.

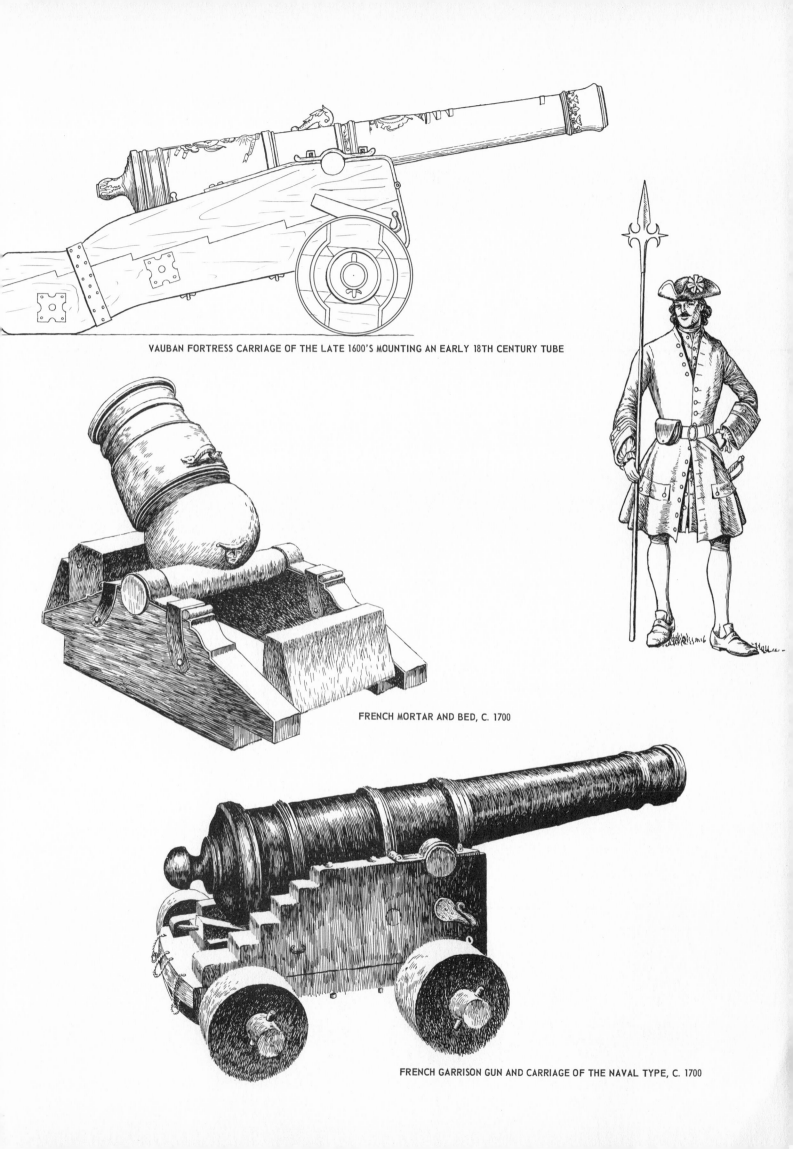

VAUBAN FORTRESS CARRIAGE OF THE LATE 1600'S MOUNTING AN EARLY 18TH CENTURY TUBE

FRENCH MORTAR AND BED, C. 1700

FRENCH GARRISON GUN AND CARRIAGE OF THE NAVAL TYPE, C. 1700

SWEDISH 4-PDR GUN ON CARRIAGE, 1756

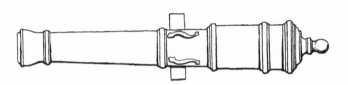

SWEDISH LIGHT 4-PDR GUN TUBE

FRENCH 4-PDR OF THE VALLIERE SYSTEM

FRENCH 8-PDR OF THE VALLIERE SYSTEM

FRENCH 8-PDR VALLIERE GUN ON CARRIAGE

consisted of a scroll bearing the name of the piece itself such as *"La Brillante"* or *"La Mutine."* Behind this was another scroll bearing the Latin motto *"Ultima Ratio Regum"*—The Last Argument of Kings. For many guns of the Vallière system the next scroll bore the name of Louis Auguste de Bouron, Duc de Maine, who was Grand Master of French artillery, and this was followed by his arms. Then, on the first reinforce, came the royal arms surmounted by a sunburst and the motto *"Nec Pluribus Impar"*—Not Unequal to Many. This had been the device and motto of Louis XIV, and it was a rather backward way of saying that the king by himself was equal to many others.

Interestingly enough, Vallière recognized no howitzers among his official artillery pieces. Perhaps this was because the French had not yet adopted the technique of placing the shell in the tube with the fuze turned away from the powder charge and allowing the flash from the charge to ignite it. With their mortars they still lit the fuze separately before applying the match to the vent to fire the piece. This was called firing at two strokes, and it was a dangerous practice. Nevertheless it was almost 1750 before the French began to "fire at one stroke" and let the flash from the charge ignite the shell fuze. Despite their lack of official recognition, however, some howitzers did appear in France and find employment by artillerymen. Since these were irregular pieces, there was no standardized size or design, but the 8-inch bore seems to have been the most popular.

Vallière did an excellent job as far as he went. The difficulty was that he stopped with the standardization of gun and mortar tubes. Carriages and beds were left to the inclinations of the various manufacturers, with the result that there were variations from one department of France to another. And there was absolutely no uniformity in the wagons and other rolling stock that supported the guns. One of the few unifying characteristics of Vallière carriages and rolling stock was their color. All were to be painted a deep red.

In all of this discussion of French artillery materiel, it might be mentioned that the French inch was slightly longer than the American inch and that the French *livre*, which is translated as pound, actually weighed a bit more than the English or American pound. Thus their pieces were a little larger than their designations might indicate. In fact, the French 8-pounder was almost exactly equal to the English 9-pounder.

The Light 4-pounder

As the years went on there were some gradual changes in the French artillery. Most of the Vallière guns had been conceived as firing from fixed positions. Even the 4-pounder was heavy for field use in direct support of infantry despite the fact that Gustavus Adolphus of Sweden had demonstrated the value of regimental pieces more than a century before. The wars of the 18th century continued to prove Gustavus right, however, and eventually the French turned to his country for a gun they could use for this purpose. Sweden had developed a light 4-pounder some years earlier, and the French artillery began to use it informally, probably in the 1740's. It was a fine little gun, short and light, and it mounted on a sharply angled carriage of its own design. Moreover this carriage featured an elevating screw with a crank handle that was well in advance of its time. In 1756 the French formally adopted the gun and began its manufacture in their own foundries.

The Gribeauval System

This was the condition of the French artillery at the time of the Seven Years War or the French and Indian War, as it is called in America. During that conflict the French came into contact with far better artillery systems and organizations, notably the Austrian artillery. As a result they began casting about for ways of improving their own setup, and to that end they called upon Jean Baptiste Gribeauval, who had served with the Austrians and who had already made a name for himself as an artillerist.

The result of this call was the famous Gribeauval System of artillery, probably the first complete system adopted by any country in the world. Gribeauval examined not only the materiel of artillery, but also its employment and the organization of the arm. And his study of artillery materiel did not stop with tubes as had Vallière's. It continued through carriages, rolling stock, equipment, and even to pontoon bridges. In each area he suggested improvements that would bring the French artillery from a dismal condition to a leading place among all European powers.

First of all, Gribeauval distinguished between cannon for field, siege, garrison, and seacoast use, and he designed them and their carriages accordingly. For field use he used 4-, 8-, and 12-pounders, and he redesigned these considerably. He did

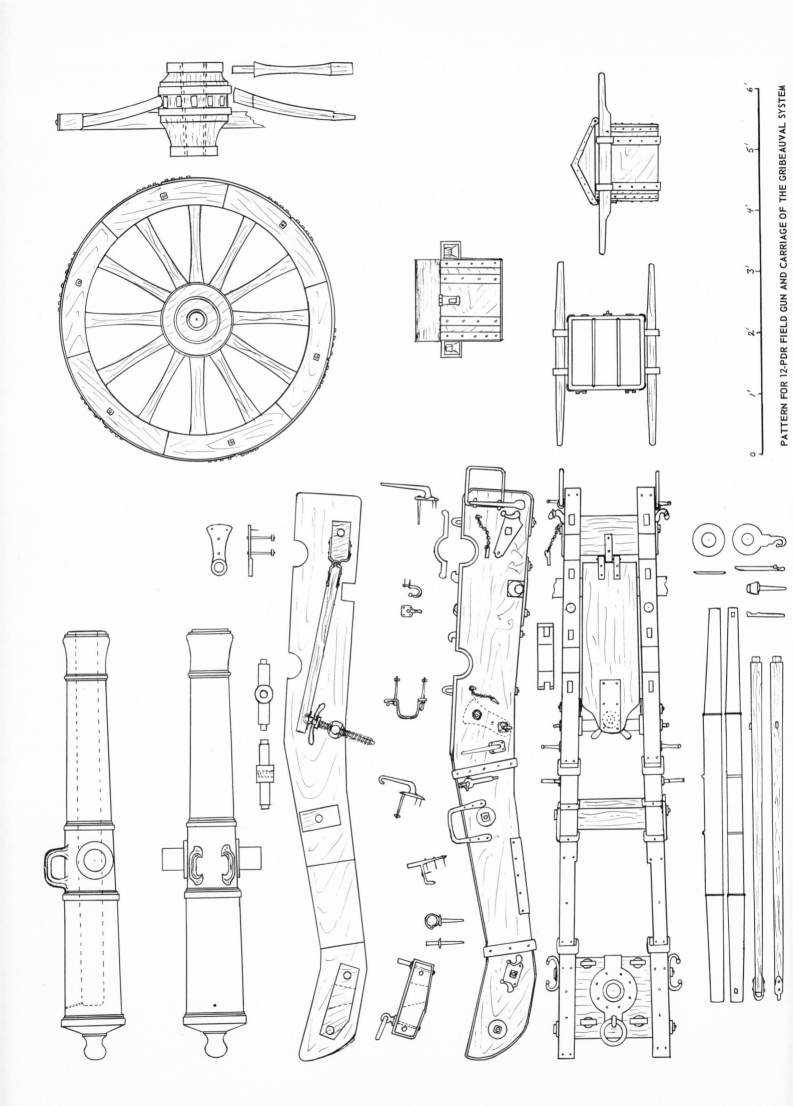

PATTERN FOR 12-PDR FIELD GUN AND CARRIAGE OF THE GRIBEAUVAL SYSTEM

0   1'   2'   3'   4'   5'   6'

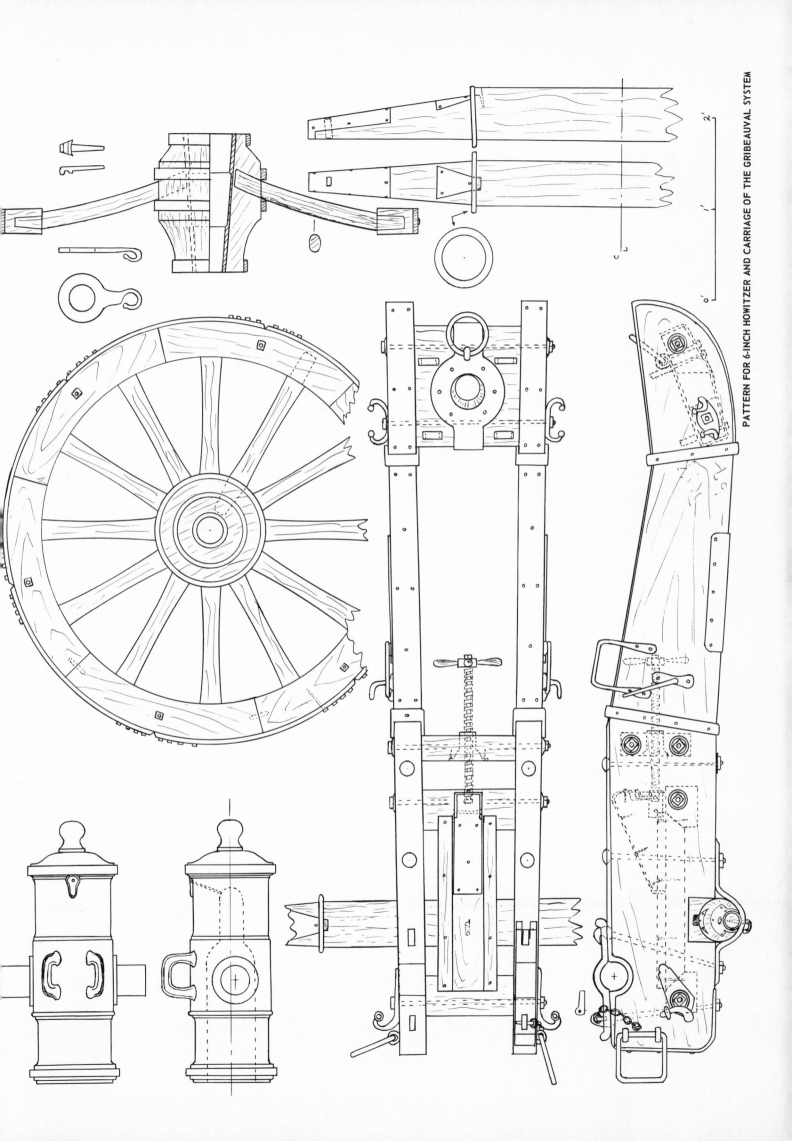

away with the cast decoration of the Vallière guns and created a smooth, streamlined appearance. Next he reduced the size of the guns, making all the field guns eighteen calibers long with a weight of fifty times the ball they threw. He retained the dolphins, but stripped them of their sculptural form so that they became simply handles. He raised the trunnions slightly so that they were just a fraction below the axis of the bore, and he added rimbases to give the gun greater stability in its carriage. He also made the howitzer an official weapon, retaining the older 8-inch size for special use and instituting a new 6-inch howitzer for field employment. On both howitzers he kept the trunnions in the old position on the low line of the bore.

His major improvements to field carriages included strengthening the members and reinforcing them heavily with iron straps in the places of greatest stress and wear. He also added an elevating screw for aiming that raised and lowered a platform on which the breech of the gun rested instead of connecting with a lug on the cascabel as the British system did. Howitzers, however, retained the wedge system of elevation. Most noticeable of all, he provided two sets of notches for the trunnions on gun carriages. One was for firing and one for traveling when the piece was limbered up so that at both times the gun would be in the best position on its mount. Finally, an ammunition box rode between the flasks ready to be lifted out for use.

For siege guns Gribeauval retained the Vallière tube designs. These were the 16-pounder and the 24-pounder. The decoration was removed, but the size and proportions remained the same, and the trunnions continued on the low line of the bore. Their carriages were strengthened and reinforced like those of the field guns, but throughout they retained the quoin for elevating. In the mortar field he kept the 8-inch model unchanged plus a 12-inch version that was seldom used and added a new 10-inch mortar. The beds of these mortars were to be made with iron cheeks and wooden cross members. They were held together by long bolts that passed through the wood, and there were knobs that stuck out on all four corners to give purchase for levers in shifting the mortar's position. These were strong durable mortar beds that did not have to be replaced like the wooden ones, and they weighed about the same as the heavy wooden blocks that had been used earlier.

The rolling stock that served these guns was just as efficient as the tubes themselves. The limber had a tongue so that the horses could be hitched abreast. Four horses were required for the 4- and 8-pounder guns and six for the 12-pounder. There was also a new ammunition wagon called a caisson. This had a box body with a peaked lid, and the inside was broken up into compartments for the efficient storage of powder charges and projectiles. Another form of Gribeauval caisson also appeared, and is known today by two surviving specimens in America. This caisson had a rounded top on the box instead of a peaked one, and it had spring suspension instead of resting the box directly on the axles. Such documentary reference as can be found seems to indicate that these modified caissons were designed for horse artillery in which the artillerists rode astride the caisson. Certainly a caisson with springs and a rounded top would have been much more comfortable for this purpose. If this is so, then the American specimens may well date from 1808, when the United States first experimented with light artillery.

All of the wooden parts of the Gribeauval rolling stock, the cannon carriages, and the ammunition boxes were painted blue until the French Revolution, when they became green. The exact shade seems to have varied, and surviving specimens which retain their paint in the Musée de l'Armée, Paris, suggest that it was sometimes a fairly light blue, sometimes almost a royal blue. The iron hardware was, of course, painted black.

Gribeauval also designed radically new and different seacoast guns and carriages, but these were not used in the United States until after 1800, and so they are described in the next chapter. In fact, it is a moot point whether any Gribeauval materiel actually saw service here during the Revolution. The system was adopted in 1767, but immediately there were protests and arguments from artillerists with other ideas. In 1772 these opponents prevailed, and the Vallière system was returned until 1774, when the Gribeauval system triumphed. None of these new Gribeauval pieces were available for sending to the American colonies during the war, since the changeover was just taking place. It may be that the French artillery that served with the American army at Yorktown had the new materiel. Indeed, most students believe they did, but the point cannot be proved. It may be that America did not see the new French artillery until after the United States had become independent.

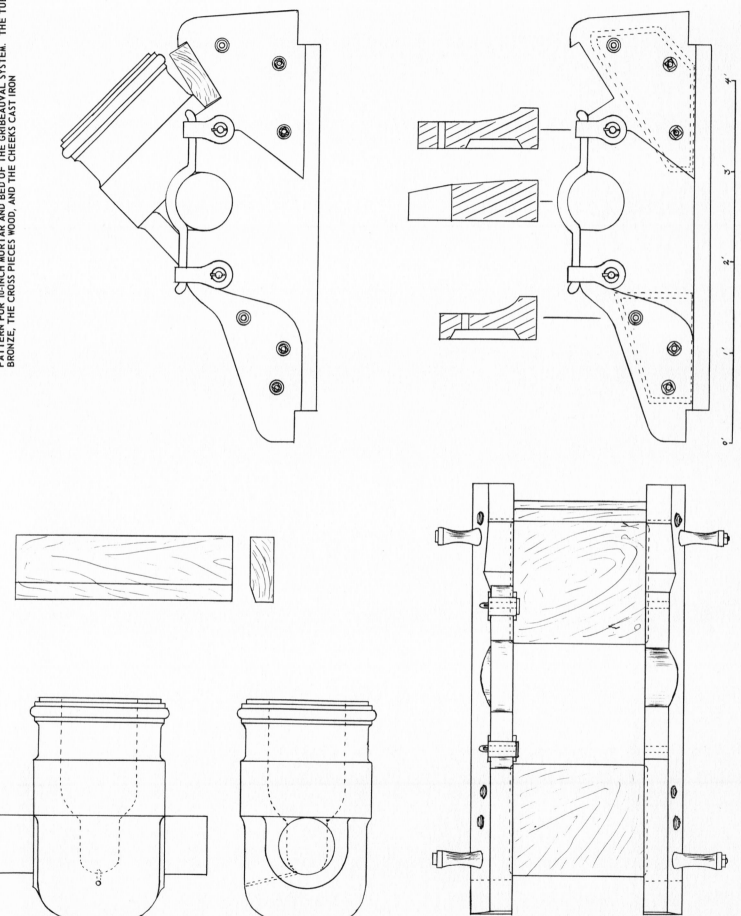

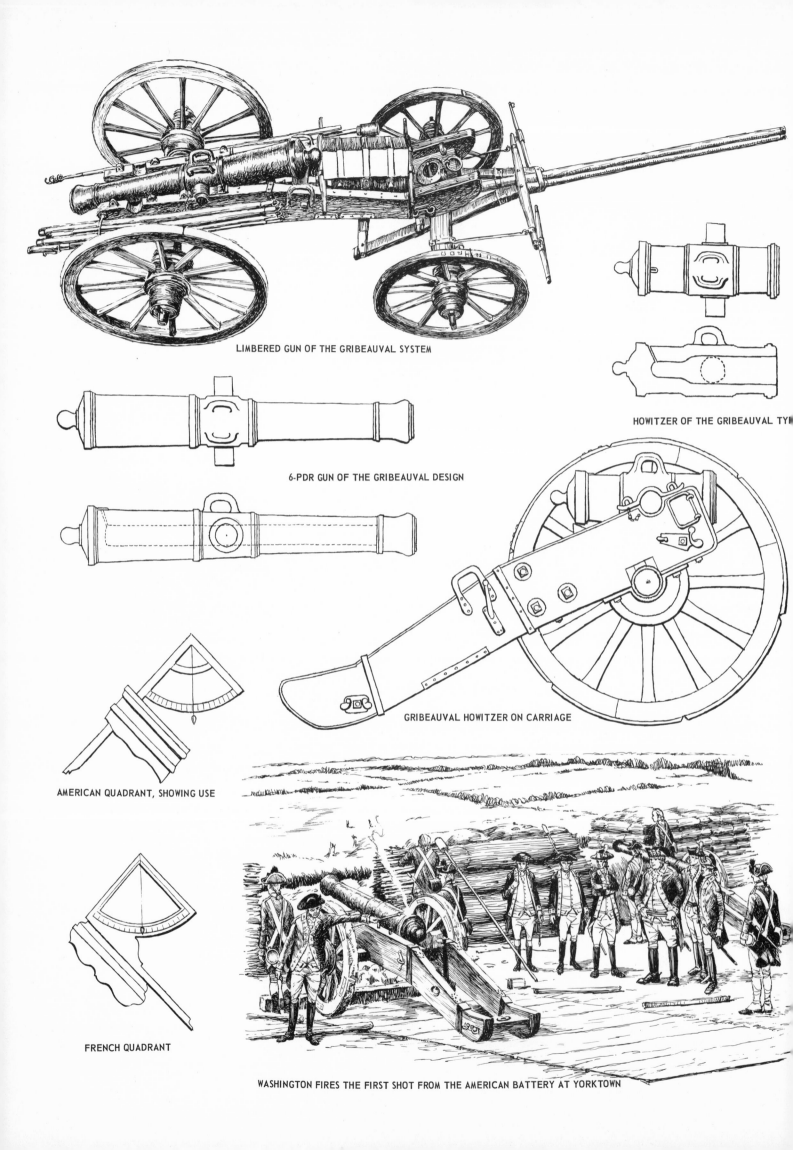

LIMBERED GUN OF THE GRIBEAUVAL SYSTEM

HOWITZER OF THE GRIBEAUVAL TY[

6-PDR GUN OF THE GRIBEAUVAL DESIGN

GRIBEAUVAL HOWITZER ON CARRIAGE

AMERICAN QUADRANT, SHOWING USE

FRENCH QUADRANT

WASHINGTON FIRES THE FIRST SHOT FROM THE AMERICAN BATTERY AT YORKTOWN

American Artillery

AMERICAN ARTILLERY came into being with the Revolution. There had been some American artillerists almost since the beginning. The Ancient and Honorable Artillery Company of Boston had spurred scientific artillery since its founding in 1638, and there had been other independent artillery companies as well. American artillerists had served with their British colleagues at the sieges of Louisburg also, but these were all local units, each one fiercely independent, and there was no uniformity either of materiel or organization. The revolution changed this. With the birth of the Continental Army in June of 1775 a national artillery arm became necessary. It started slowly and inefficiently, but with the appointment of Henry Knox as colonel of the Continental Regiment of Artillery on November 17, 1775, things began to improve. A genius at organization and training, Knox eventually built the Continental artillery up to four regiments and trained them to equal any artillerymen in the world.

A Mixture of Materiel

Still, there was little that Knox could do about materiel. The Americans had to take what they could get. They were able to cast some guns themselves, but the rest of their train of artillery consisted of pieces acquired from France or captured from the enemy, found in colonial arsenals, or taken from ships. Naturally this led to a great multiplicity of types and sizes. All of the British and many of the French sizes were in use, and since the French sizes differed from the British, they either had to be segregated or rebored to British calibers. Some idea of the complexity of an American artillery train can be gained from Knox's estimates of the needs for the campaign of 1778:

Brigade artillery, seventeen brigades, with four guns each:—sixty-eight pieces to be 3, 4, or 6-pounders; with the park—two 24-pounders, four 12-pounders, four 8-inch howitzers, eight 5½-inch howitzers, ten 3 or 4-pounders, ten 6-pounders; for the reserve,—to be kept at a proper distance from camp,—thirty 3, 4, and 6-pounders, two 12-pounders, one 24-pounder; all the foregoing brigade, park, and reserve guns and howitzers *to be brass.* In addition, twelve 18-pounders, twelve 12-pounders, battering pieces, on travelling carriages, together with two 5½-inch and twelve 8, 9, and 10-inch mortars; the battering pieces and mortars to be *cast iron.*

Among those cannon that America received from France were some of the light Swedish type 4-pounders that the French Army was using for regimental pieces. Thirty-one of them, for instance, arrived with their carriages on the ship *Amphitrite* early in 1777, and Knox welcomed them immediately. He was less pleased with the other guns of the Vallière system which were on board. These were too big and heavy for his purposes, he thought, even the 4-pounders. Each of these 4-pounders, Knox estimated, would contain nearly enough metal to cast three light 6-pounders sufficient for service, and he asked that the French guns be melted down and recast in this fashion. The carriages could be cannibalized for parts.

Cannon Made in America

When Americans did make artillery of their own, they followed British patterns. As a matter of fact, John Muller's *A Treatise of Artillery* appeared in a pirated edition in Philadelphia in 1779, dedicated to "George Washington, General Henry Knox and Officers of the Continental Artillery." It was the only technical treatise on artillery available to American artificers, and so they followed it. Thus American guns were apt to be fourteen calibers long with trunnions on the center line of the bore as Muller recommended, even though British guns held to the older designs. Aside from Muller's influence, British designs and theory predominated because these were what Americans had become familiar with in the years before the war. Naturally they clung to them. Also

GRIBEAUVAL CAISSON

GRIBEAUVAL CANNON LIMBERED UP

GRIBEAUVAL CAISSON, POSSIBLY FOR HORSE ARTILLERY

AMERICAN GUN CREW FROM A CONTEMPORARY POWDER HORN ENGRAVING

MONOGRAM FROM A REVOLUTIONARY WAR HOWITZER

MONOGRAM FROM A REVOLUTIONARY WAR HOWITZER

BRASS HOWITZER CAST BY N. BYERS

J. Byers
1777 Fecit
Philad͞a

BYERS' SIGNATURE FROM A TYPICAL CASTING

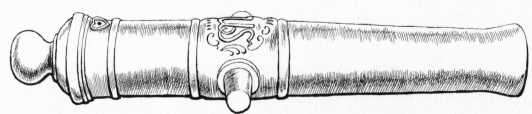

GUN CAST BY N. BYERS

they were better designs than the Vallière system, and news of the Gribeauval types had not yet reached America in detail.

Some of these tubes made in America were simple and even crude. Others were quite handsome. Some of the brass pieces bore decorations just as their European counterparts did, but the liberty cap on a pole and the sunburst replaced the royal cypher or royal arms. The letters "US" and "UC" (for United Colonies) also appear on some guns and so do the maker's name, the place, and even the date of manufacture.

Carriages and Colors

As in the British service, oak was preferred for building carriages in America. When they could not obtain oak, however, Continental artificers used other hardwoods including walnut and chestnut. Elm and beech were recommended for wheels, but hickory appeared on occasion as did almost any other tough wood when necessity required it.

When it came to color, however, American carriages seem to have offered a variety of hues unparalleled by any other nation. The iron gun tubes were almost always painted black, and the hardware of the carriages also was painted black as a general rule. In some instances, though, the mountings seem to have received only a coat of red lead. This may have been caused by a paint shortage, for it was the wooden elements that offered the real variety of colors. At the beginning many of them seem to have been painted gray like those of the British artillery. In 1776, for instance, General Philip Schuyler ordered sufficient mate-

rials "to paint 250 carriages of a lead colour. . . ." The Charles Willson Peale portrait of Washington at Princeton shows him leaning against a gray gun carriage, and the paintings of the battle of Princeton by James Peale and William Mercer, done shortly after the event, show a similar gray carriage. At the same time, however, these paintings also illustrate a red brown carriage. This may have been done to indicate an imported Vallière system carriage or it might be an American carriage, for there is some documentary evidence to show that the Continental Artillery sometimes painted its carriages that color, too. It might even indicate a carriage that had not been painted at all, but merely oiled and allowed to weather. By 1780 there are references to painting carriages blue, perhaps under the influence of the French, who had changed to that color with the Gribeauval system, and before the end of the war in 1783 blue seems to have become standard for all American carriages.

Whatever the color of the carriage or the origin of the tube, American artillerymen learned to serve them well. The field guns performed with American crews from Bunker Hill till the end of the war. Garrison guns lined the walls of fortresses like Ticonderoga and West Point and occasionally put to sea as did those which Benedict Arnold mounted on the vessels of his Lake Champlain fleet. Siege pieces had perhaps the longest service of all, stretching from the encirclement of Boston in 1775 till the capture of Yorktown in 1781. In fact, it was these siege guns, accompanied by the French artillery, that made Yorktown untenable for the British and thus brought an end to the Revolution.

Spanish Artillery

SPANISH ARTILLERY of the period closely resembled the French—with a little touch of the English as well. The brass guns of the early period clung to 17th century forms with bulbous multi-ringed muzzles, and they retained this characteristic at least until 1770. Dolphins were ornamented sculpturally, and like the French pieces, these guns frequently bore individual names on a scroll just behind the muzzle. Another characteristic that usually seems to apply lies in their proportions. If a piece were divided into six equal parts from the base ring to the rearmost of the muzzle moldings, the first reinforce would equal two parts, the second reinforce would equal one part, and the chase would correspond to the remaining three. This arrangement with a long first reinforce, a short sec-

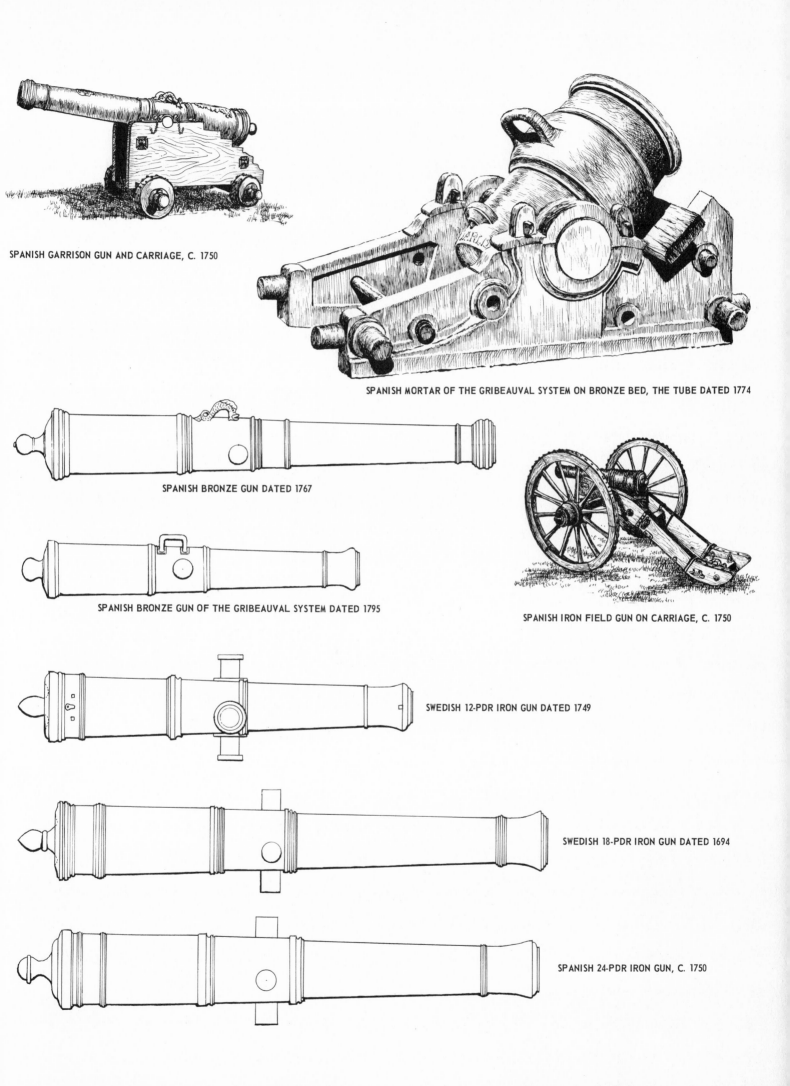

SPANISH GARRISON GUN AND CARRIAGE, C. 1750

SPANISH MORTAR OF THE GRIBEAUVAL SYSTEM ON BRONZE BED, THE TUBE DATED 1774

SPANISH BRONZE GUN DATED 1767

SPANISH BRONZE GUN OF THE GRIBEAUVAL SYSTEM DATED 1795

SPANISH IRON FIELD GUN ON CARRIAGE, C. 1750

SWEDISH 12-PDR IRON GUN DATED 1749

SWEDISH 18-PDR IRON GUN DATED 1694

SPANISH 24-PDR IRON GUN, C. 1750

ond one, and a chase only as long as the two reinforces together gives a very characteristic silhouette to these Spanish guns.

At about the same time there were also iron guns. At such forts as the Castillo de San Marcos and elsewhere among Spain's former colonies there are still encountered iron guns that look almost English, and others have been recovered from Spanish shipwrecks. These guns sometimes have belted cascabel buttons, trunnions in the low line, and general English silhouettes, but the mouldings are just slightly un-English. Such iron guns exist in both field and garrison sizes, and two are illustrated. On the relationship between England and Spain in cannon design, it might be mentioned that Muller's *Treatise of Artillery* was translated into Spanish and seems to have been influential.

The relationship between the ruling houses of Spain and France had kept those two nations close since 1700, and French influence on Spanish artillery design was notable. Spain never did adopt the Vallière system but quickly accepted the Gribeauval. It is interesting to note, however, that surviving Spanish mortar beds of the Gribeauval pattern are made with bronze cheeks instead of cast iron as called for by the French. One of these beds, lacking its wooden rear cross member, is shown.

Spanish Carriages

Spanish gun carriages in the pre-Gribeauval period varied somewhat according to the builder. Field carriages were split or flask trail design usually with only a slight angle on the flasks. They were quoin-elevated. Garrison carriages tended to be a little taller than English or French models, and the Spanish had a fondness for large iron washers in the form of Maltese crosses under the transverse bolt heads. When it comes to the color they were painted there is room for speculation. Some may have been painted blue since this was a Bourbon color and hence appropriate to the ruling house of Spain. No records confirming this have been found, however. In fact, the only comments on treating the wooden members of gun carriages state that they are to be coated with turpentine. This seems to indicate that at least in some instances the Spanish carriages were simply oil-treated and left to weather naturally.

Spanish artillery had been the first to come to America, but its influence had rapidly waned. During the 18th century only the forts in Florida mounted any considerable cannon. These posts were lost to Great Britain at the end of the French and Indian War in 1763, but regained at the close of the American Revolution twenty years later. When Spain reoccupied her forts at that time she probably filled out their armaments with Gribeauval guns and mortars, though she may have transferred older pieces from some of the depots in the Caribbean, for Florida was, after all, still an insignificant outpost. Undoubtedly there were Spanish Gribeauval pieces in Florida by the time Spain ceded that territory to the United States in 1821, and aside from some small pieces that may have found their way into the outposts in California or the Southwest, these were the last Spanish artillery in the continental United States.

Swedish Artillery

SWEDEN HAD manned no colonies in the United States since the ill-fated posts along the Delaware River in the early 1600's. Nevertheless Swedish cannon played a part in the history of American artillery right down to the close of the American Revolution. Ever since Gustavus Adolphus, Sweden had been a pioneer in cannon design and manufacture. Her guns were among the best in the world, and the adoption of the light Swedish 4-pounders by the French and their eventual use by the Continental Army have already been mentioned. More than that, however, Sweden was an important supplier of cannon to many of the nations of Europe, especially guns for arming ships. Most of these guns were iron, and a surprising number of them are still to be found in the United States today, where they have remained after being removed from their vessels. Some found their way into fortress armament, and it is interesting to note that two of the three iron guns on the *Philadelphia,* Arnold's flagship in the Continental Army fleet on Lake Champlain, are Swedish. These guns on their carriages were recovered from the bottom of the lake and are now exhibited in the Smithsonian Institution. Other Swedish guns may be seen at the Castillo de San Marcos at St. Augustine, and in other American cities.

Since some of these guns were cast commercially rather than as part of a national supply, they naturally vary somewhat. Nevertheless there are certain characteristics that make the iron guns immediately recognizable. Most noticeable is the breech face, which is flat or slightly concave. Second is the cascabel button, which is usually shaped somewhat like a pine cone; and third are the mouldings, which often look as if someone had wound a tight, smooth rope around the tube. Otherwise the lines are reminiscent of the British. The trunnions are on the low line of the bore, and a few of them have rimbases and a corresponding rim around the external end. These guns are seldom marked in any way, and it is often a moot point whether they are actually Swedish or Danish. For this reason some students prefer to call them simply Scandinavian. Whichever they may be, they were plentiful and a definite factor in the history of American artillery.

Advances in Ammunition and Firing Technique

THE AMMUNITION of the French wars and the Revolution closely resembled that of the previous period. Gunpowder changed not at all. Some of the more wildly imaginative projectiles disappeared, but the solid ball, the shell, the cross bar shot, grape, cannister, and carcasses remained. Some British iron cannon balls were made with indented broad arrows cast into them, and the fragment of one French shell with a raised fleur-de-lys has been found at Yorktown, but in most instances the cannon ball and the other projectiles remained much as they had been.

Sabots and Cartridges

The most notable innovation involving the cannon ball was the introduction of the sabot. This was a wooden disc hollowed out on one side to receive the base of the ball, which was attached to it by two straps of tinned iron. This sabot served in place of a wad and also as a means for attaching the powder bag so that an efficient form of fixed ammunition for solid shot could be manufactured and used for field guns. Similar fixed rounds for cannister and grape were also used. These fixed rounds were used only for rapid firing, however; for most shooting in field guns and for siege and garrison pieces they employed semi-fixed loads with a separate projectile and a bag of powder.

The best of these powder bags or cartridges were normally made of flannel. Paper and parchment cartridges were still in use also, but both had drawbacks. The paper ends of the bags did not usually burn completely, and so a wad of paper gradually built up at the base of the bore until it blocked the vent and had to be removed with a worm or wad hook. Also it smoldered and so posed a danger of setting off the next charge prematurely if the sponge did not put it out. Parchment shriveled up with the heat when the charge ignited, and sometimes it found its way up into the vent, where it had to be cleaned out with the vent pick or priming wire. Flannel, on the other hand, burned cleanly and was not so apt to leave smoldering remnants. Thus artillerists greatly preferred the cloth cartridges when they could get them.

Firing at One Stroke

Shells continued in use for both mortars and howitzers. As mentioned previously, artillerists of this period learned that the flames from the burning powder charge would pass around the shell and ignite the fuze, even though this was on the side of the shell facing the muzzle. This was a great discovery for artillery. When the fuze had been placed facing the charge, the force of the explosion had frequently driven it right into the powder cavity and caused a premature burst. Thus most artillerists had placed the shell in the bore with the fuze pointing forwards, lit it separately with their linstock and then fired the cannon. This was called firing at two strokes, and it was both dangerous and inconvenient. If the piece failed to fire there would be a shellburst inside the bore which well might ruin the mortar and injure the crew. Also such firing was only practical for a piece with a short bore such as a mortar or possibly a howitzer. When the practice of firing and letting the flash ignite the fuze was discovered it meant that shells could also be used for guns with their longer tubes. Gunners did not seem to recognize the significance of this possibility immediately, however. They conducted some experiments in shell firing from guns, but did not make it a regular practice.

Hot Shot

One special firing technique that came to the fore during this era was the red-hot shot designed to set fire to an enemy ship or building. These incendiary projectiles could be fired from gun, howitzer, or mortar with equal facility, and they were effective. It was a red-hot shot from the American siege battery at Yorktown which set the British frigate *Charon* afire and sank her in 1781, and there are numerous other examples of their effectiveness. To fire hot shot, the gun crew first prepared a hole in the ground about six feet in diameter and four feet deep. In this they built a hot fire, then placed the balls on top. Then they prepared the artillery piece with a load of powder, a wad, and either a wooden disc the exact size of the bore or a thick piece of sod to keep the hot ball from igniting the powder when it was seated.

They tried to have everything ready so that the shot could be fired as soon as they placed the ball in position. Otherwise the shot would cool and lose its effectiveness. It was, after all, a dangerous and tedious technique, and there was no sense in doing it if the shot was going to arrive at its target too cool to set a fire.

Portfires

There were also significant advances in the devices for priming and firing cannon. The old slow match continued to be used, but by the mid-18th century it was more often used for lighting a new device called a portfire that the artillerists then employed to fire the cannon itself. These portfires contained a quick-burning composition of gunpowder, saltpeter, and sulphur moistened with linseed oil, and they burned like a flare with a very hot flame. Holding this portfire in a stick called a portfire stock, the gunner applied the flame to the priming in the vent of the piece. This priming might be loose powder. Indeed loose powder was the usual priming through most of the period, and it continued in use well into the early 1800's. The artillerist poured this powder into the little pan or vent field that surrounded the vent, and William Stevens, who had served in the Continental Artillery, went so far as to recommend that he then take his powder horn and roll it over the powder, crushing the grains and supposedly making it quicker to ignite. It seems a somewhat dubious practice at best, and in an action there would not have been time for it anyway.

Priming Tubes

Sometimes, instead of loose powder the gunners used priming tubes. These were tubes of tin or reeds filled with a quick match or portfire composition. One end was flattened into a dish shape to give a greater area for receiving the flame from the portfire, and a piece of paper was usually pasted over this end to keep the powder in place and prevent it from drying completely. Tin tubes often had the other end sharpened to pierce the cartridge bag. These primers averaged from 4 to $6\frac{1}{2}$ inches in length for most field guns. No one knows when or where these primers first appeared, but they were definitely in use in both the American and British artillery during the Revolution. Most manuals of the period recommended them

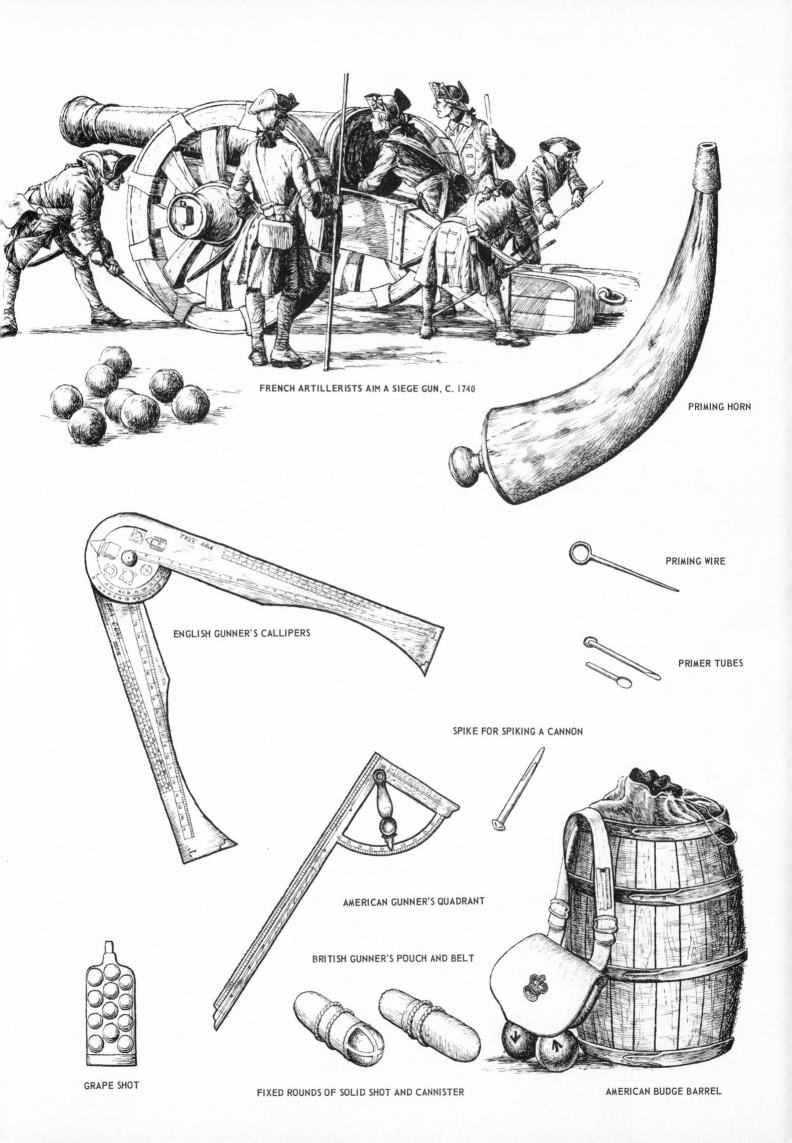

FRENCH ARTILLERISTS AIM A SIEGE GUN, C. 1740

PRIMING HORN

ENGLISH GUNNER'S CALLIPERS

PRIMING WIRE

PRIMER TUBES

SPIKE FOR SPIKING A CANNON

AMERICAN GUNNER'S QUADRANT

BRITISH GUNNER'S POUCH AND BELT

GRAPE SHOT

FIXED ROUNDS OF SOLID SHOT AND CANNISTER

AMERICAN BUDGE BARREL

only for quick firing, leaving the loose powder priming for more leisurely shots. The idea was to conserve the more expensive primers for times of real need. Then, as now, the administrators were economy-minded, but the cannoneer in the field probably paid little heed. He undoubtedly used his primers and his fixed ammunition any time he could find a supply of them. Cannoneers, after all, have changed less than their ammunition.

The big changes in loading and firing cannon during the 18th century lay in technique. The sighting instruments might be a little more sophisticated, but they remained basically the same, and there were few notable changes in the implements. True, the French did come up with a bent-handled sponge for their Swedish light 4-pounder which was unlike anything used previously, and a straight linstock with a hole also appeared, but these were minor variations. When it came to employing the tools, however, there was a major revolution.

The Gun Crew

This was the development of the gun crew. Instead of one or two men and some unskilled pioneers, a trained group of skilled artillerists loaded and fired the pieces with a speed unheard of in previous centuries. Each man had a specific task and drilled with his fellows to perform it at exactly the right time and in the safest and most efficient manner. There are records that indicate good British and German crews could load and fire an astounding 12 or 14 unaimed shots a minute. Two competing crews did just that in 1777 as they prepared to invade upper New York State with Burgoyne's army, and Louis Tousard, writing in 1809, indicated that the French artillery could match it.

In the British and American Service at the time of the Revolution, a full gun crew for a 6-pounder field piece consisted of 14 or 15 men. Each of them was numbered. The first six of these were the least skilled. They manned the drag ropes that maneuvered the piece. When prepared for action, they stood three on either side of the carriage opposite the wheels. Number 7 stood at the right of the muzzle with the sponge and rammer to push home the shot and to sponge out the bore between shots. Number 8 stood at the opposite side of the muzzle and actually placed the ammunition in the bore. Number 9 stood at the right of the breech and thumbed the vent while number 7 sponged and

rammed. This prevented smoldering pieces of cartridge bags from being forced up into the vent, and it also helped create a suction when the sponge was withdrawn that would help extinguish any sparks that might have survived the sponging. He also probably primed the vent after the charge had been rammed home. Number 10, on the left of the breech, held the lighted portfire and actually fired the piece by reaching over and applying the lighted end of the flaming compound to the primer in the vent. Both numbers 9 and 10 stood outside the wheels when the piece was touched off to be clear of the recoil. Numbers 7 and 8, however, remained inside the wheels, close to the muzzle, their mouths open to protect their ear drums from the force of the blast at such close range. Number 11 manned a handspike that fitted into a socket in the trail transom. With it he pointed the gun as directed by the officer in charge. Number 12 stood clear of the trail transom on the right and held the water bucket, the linstock with its lighted wick, and a spare portfire in a stock ready to supply number 10 upon demand. Number 13 was a runner who carried ammunition from number 14's supply and handed it to number 8 for inserting in the muzzle. Number 15 also presided over a supply of ammunition and at the same time he held the limber horse. The civilian driver, who did not count in the crew, had meantime unhitched the lead horse and led him around to the rear.

This was the ideal crew. In action the functions were often carried on with fewer men as casualties depleted the number available. Each man, therefore, was supposed to know the duties of every other member of the crew and to be able to perform them if necessary. Sometimes infantrymen had to be pressed into service for the less demanding tasks, and in the American artillery even women served on gun crews in emergencies. Margaret Corbin and Mary Ludwig Hays, known as "Molly Pitcher," both helped on gun crews, the latter serving as number 13, a position that contemporary manuals suggested be given to an especially strong man because of the strenuous running back and forth with ammunition.

Field guns like the 6-pounder required the biggest crews. Siege and garrison pieces that were too heavy to be maneuvered by drag ropes discarded the first six men and operated with eight or nine men. Mortars employed fewer still, though the men handling the ammunition had the additional task of setting fuzes for the shells.

PORTFIRE AND ENGLISH STYLE STOCK

LINSTOCK WITH HOLE

AMERICAN GUN CREW FROM JAMES PEALE'S CONTEMPORARY PAINTING OF THE BATTLE OF PRINCETON

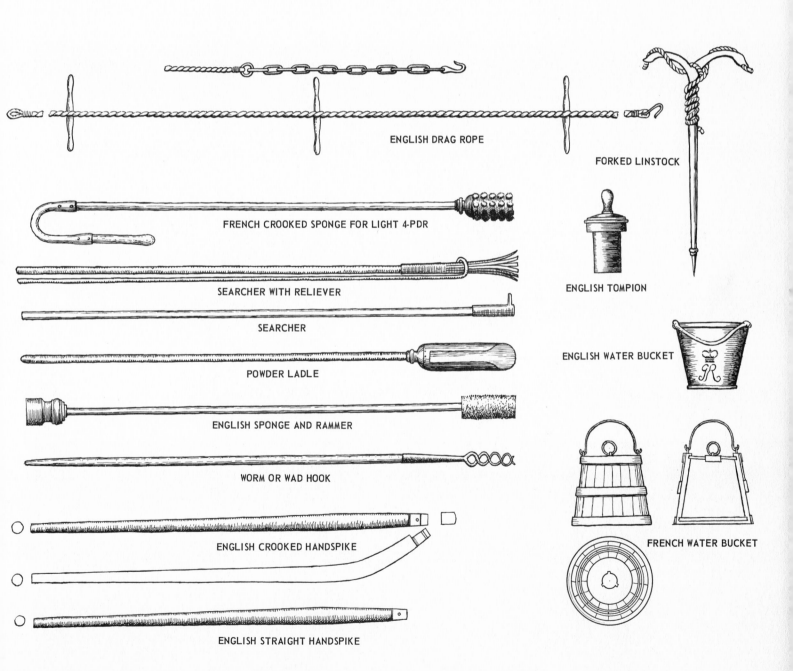

ENGLISH DRAG ROPE

FORKED LINSTOCK

FRENCH CROOKED SPONGE FOR LIGHT 4-PDR

SEARCHER WITH RELIEVER

SEARCHER

ENGLISH TOMPION

POWDER LADLE

ENGLISH WATER BUCKET

ENGLISH SPONGE AND RAMMER

WORM OR WAD HOOK

ENGLISH CROOKED HANDSPIKE

FRENCH WATER BUCKET

ENGLISH STRAIGHT HANDSPIKE

Spiking Cannon

IN ADDITION to the gun crew one other technique seems to have developed during this period. From these years come the earliest references to spiking guns to put them out of action and render them useless to an enemy. This was done when guns had to be abandoned or when a battery was taken briefly with no intention of holding it and no means for hauling off the guns. To spike a gun an artillerist drove a rod through the vent and tried to bend it inside so that it could not readily be removed. Almost any rod of the proper diameter would do. There are records of men using their bayonets for the purpose. But there were also special spikes carried by gunners with split tips that would spread out like a cotter pin when they struck the bottom of the bore. Such spikes were carried at least by British artillerymen during the Revolution, and they were probably employed by other nations as well. Other spikes had soft iron tips that would bend easily against the bottom of the bore or which could be clinched by driving the rammer down against them.

There were other ways of putting a gun out of action as well. One method involved knocking off one of the trunnions with a sledge hammer, but a more common way was to wedge a cannon ball in the bore using wooden wedges that could be driven home with the rammer. To be effective this ball had to be tight against the base of the bore. Otherwise a repair crew could sift powder down the vent and possibly blow it out. With a properly placed ball the only way to remove it was to try to burn the wedges, a difficult and time-consuming task.

The Art of Gunnery

IN AIMING a gun artillerists had the advantage of much more scientific information than their predecessors. Benjamin Robbins had brought ballistics into the modern age with his *New Principles of Gunnery* in 1742. Still a man could not aim farther than he could see, and firing at long range was still accurately called "firing at random." Windage was too great for accurate shooting much over 500 yards, and projectiles drifted unpredictably to right or left because of flaws in casting. The 18th century gunner had better calipers and better quadrants than his predecessors, but his guns were no more accurate, and so the new tools did not help a great deal. Many an old gunner ignored such baubles completely. Gunnery, after all, was still an art. It took months or even years to learn the capabilities and idiosyncrasies of any individual gun or class of guns whether they were British, French, Spanish, American, or Swedish. Experience was the only thing that really counted. It was the artillerist's chief stock in trade. With enough of it he could make any of the pieces of the period perform to their utmost.

CHAPTER III

The New Nation, 1784-1835

The new nation suffered growing pains. In the artillery this was reflected by a country-wide suspicion of standing military establishments, especially those which might be considered offensive in nature, by almost continuous experimentation to devise cheaper materiel made as much as possible of native materials, and by everlasting economies. Artillery regiments were organized, but often only one company had either guns or horses; the rest served as infantry or dragoons on occasion, and even the companies with guns sometimes found themselves fighting with muskets or pistols as did the men of Washington's Battery B during the Seminole War. An ambitious system of coastal fortifications was built, but frequently there was only one company of artillerymen to man the guns. And sometimes only militia artillerists were available to do this.

Still the artillery fought widely, and sometimes spectacularly. A train of artillery accompanied "Mad" Anthony Wayne's troops against the Indians in Ohio, and two little howitzers actually opened the battle of Fallen Timbers in 1794. A company of regular artillery assisted by militia and seamen from Commodore Rodgers' fleet helped hold Fort McHenry against the British attacking force in 1814 with great gallantry. And in the last battle of the War of 1812 American artillery proved decisive at New Orleans. Rifle enthusiasts have long claimed that battle as a victory for rifles, and indeed the rifles that were there apparently performed very well indeed. Artillerists, however, note the eight batteries firing grape and cannister all along the American line plus an enfilading fire from the river, and they cheer the comments of the British veteran Gleig, who speaks of the mangled condition of the corpses of the men who died in the British assault. This could only have been the work of artillery. It was a motley group of gunners—some regular, some militia, some pirates from Barataria, and some seamen—but they did their job well, and the victory at New Orleans is as much theirs as it is the rifle's.

This was the era of the elite militia batteries. States from Massachusetts to Georgia witnessed the burgeoning of independent artillery organizations. Older units like the Ancient and Honorable Artil-lery Company of Boston, the Newport Artillery, and the United Train of Artillery in Rhode Island were joined by the New York Battalion of Artillery, The Jackson Artillerists of Philadelphia, the La Fayette Artillery of Richmond, the Norfolk Light Artillery Blues, the Portsmouth Light Artillery, later called Grimes' Battery, the Chatham Artillery of Savannah, and many others, some of which remain active today. These organizations were a vital part of the American artillery, for there were never enough regular artillerists to man coastal forts, guard the frontiers, or accompany troops into the field in time of war. At the same time they brought a color and variety to the artillery that has never been matched since.

Artillery Innovations

EXPERIMENTATION CONTINUED throughout the period also. At the beginning most gun tubes were made of brass, but copper was scarce in America and tin was unknown, while iron was common. Also iron was cheaper. Thus ordnance experts decided to try iron for guns, and they did so almost exclusively after 1800. They varied the form of guns, too, following the French designs for a while, then simplifying even those—and they tried iron carriages as well. In 1808 they actually put a horse artillery company into the field commanded by Capt. George Peter. It was a fine idea, but it lasted only a year before economies unhorsed it. There was a gradual shift from English to French patterns in carriages, and a simplification of gun tubes. A new type of cannon, the so-called columbiad, was supposedly invented, and the first of a whole series of ordnance boards were appointed to make a thorough study of European materiel prior to the adoption of a complete American system at the beginning of the next period.

These were exciting years for the artillery of the new nation and they were frustrating years as well, but they were a necessary prelude to the great blossoming of American muzzle-loading artillery that came directly after. They laid the groundwork on which eventual success was built.

Field Artillery

FROM THE very beginning of the new United States Army there were field artillery-men. When almost all the rest of the Army was disbanded at the end of the Revolution, one battalion of artillery remained. As the years passed new reorganizations of the Army found the field artillerymen serving as parts of the legions, as a Corps of Artillerists and Engineers, as two regiments of artillerists and engineers, as part of a Regiment of Artil-

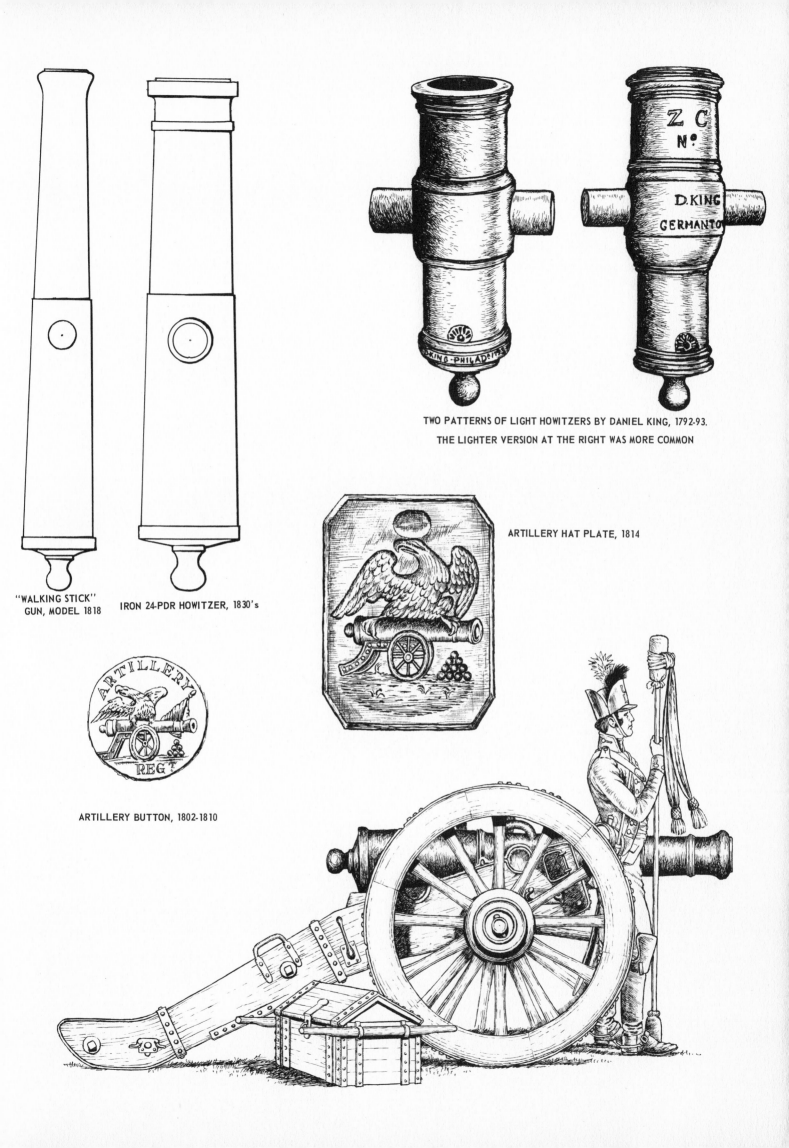

TWO PATTERNS OF LIGHT HOWITZERS BY DANIEL KING, 1792-93.
THE LIGHTER VERSION AT THE RIGHT WAS MORE COMMON

ARTILLERY HAT PLATE, 1814

"WALKING STICK"
GUN, MODEL 1818

IRON 24-PDR HOWITZER, 1830's

ARTILLERY BUTTON, 1802-1810

lerists alone, and finally as a Regiment of Light Artillery. With them were the various militia batteries, most of whom were also field artillerymen.

At first these early American artillerists clung to the British style materiel they had known since before the Revolution. They continued to use cannon that had fought through that conflict, and soon they began to cast pieces of their own following British models and using brass as the principal metal. The United States had a federal gun foundry at Springfield, Massachusetts, though reports seem to indicate that its products were not the very best. Paul Revere made better guns in his foundry, and further south J. Byers of Philadelphia continued to make fine cannon. He was joined by Daniel King of Germantown, and there were others.

The King Howitzers

Some of King's pieces, in fact, opened the Battle of Fallen Timbers, the country's most important land engagement of the 1790's. These little pieces were howitzers with 2¾-inch bores—just about big enough to throw a grenade. They were made in two styles, one somewhat heavier than the other, though both were extremely light weapons, admirably suited to accompany an army into the wilderness. The smaller version was 16 inches long and weighed 38 pounds; the heavier one measured 17 inches and weighed 60 pounds. No one knows how many of these howitzers King actually made, but nine of them still survive. As such they are among the very few American-made pieces that can definitely be dated to this early period.

General "Mad Anthony" Wayne first received the little howitzers at Pittsburgh in 1792. Apparently they were the lighter of the two models, and the trunnions proved too weak to withstand the pressures of firing. Thus Wayne asked for a heavier version weighing 60 or so pounds. This weight, he estimated, would bring the total weight with carriage up to about 212 or 224 pounds, not too much for a good pack horse to carry and it would be a much stronger piece than the 38-pound version. The new howitzers arrived in time, and they were provided with solid 3-pound shot, shell and cannister. And these were the weapons, under the command of Lt. Percy Pope, which opened the Battle of Fallen Timbers by firing shell and cannister on the hostile Indians. They played a significant role in that victory, and they continued in use along the frontier. Thirty-nine of them

appear on the records of various military posts in 1802, and some may still have been in service during the War of 1812.

Conversion from Brass to Iron

The 60-pound howitzers were among the last American field pieces to be cast in brass for more than a quarter of a century. Henry Dearborn became Secretary of War in 1801, and one of his first actions was to convert the founding of American cannon from brass to iron. He had two principal reasons for this move. One was to save money, and the other was to utilize native materials for military purposes. Iron was much cheaper than brass, and America produced fine-quality iron ore, while most brass had to be imported. He met considerable opposition in ordering iron cannon, but he persisted, and the fine record of American iron tubes in the War of 1812 attests to his wisdom.

Streamlining of Guns

These iron guns still resembled the English designs in many respects, though they did not usually have belted cascabel knobs. Americans were already looking to the French for many phases of artillery materiel but not for tube design. They retained the chase girdle and at least a ring around the breech ahead of the vent field. Also they kept to the old British series of calibers: 6-, 12-, 18-, 24-, and 32-pounders instead of the French 4's, 8's and 12's. Gradually, however, some changes began to appear. By 1812 or 1813 the second reinforce was gone and with it the chase girdle as the guns assumed a much more streamlined effect.

Then the process went too far. The guns of 1812 were still substantial and strong. In 1818 an ordnance board adopted a new model gun that was so slender it quickly acquired the nickname of the "walking stick" model. All mouldings but the base ring vanished, and the walls became too thin for practicality. They performed so badly and burst so often that they gave all iron guns a bad name despite the fine performance of Dearborn's designs. Some iron guns of improved pattern were built experimentally during the 1820's and early 1830's, but production was limited as various boards studied the subject in preparation for the return to brass or bronze in 1836.

There were, of course, a few brass guns made after 1801, and the militia could always do what

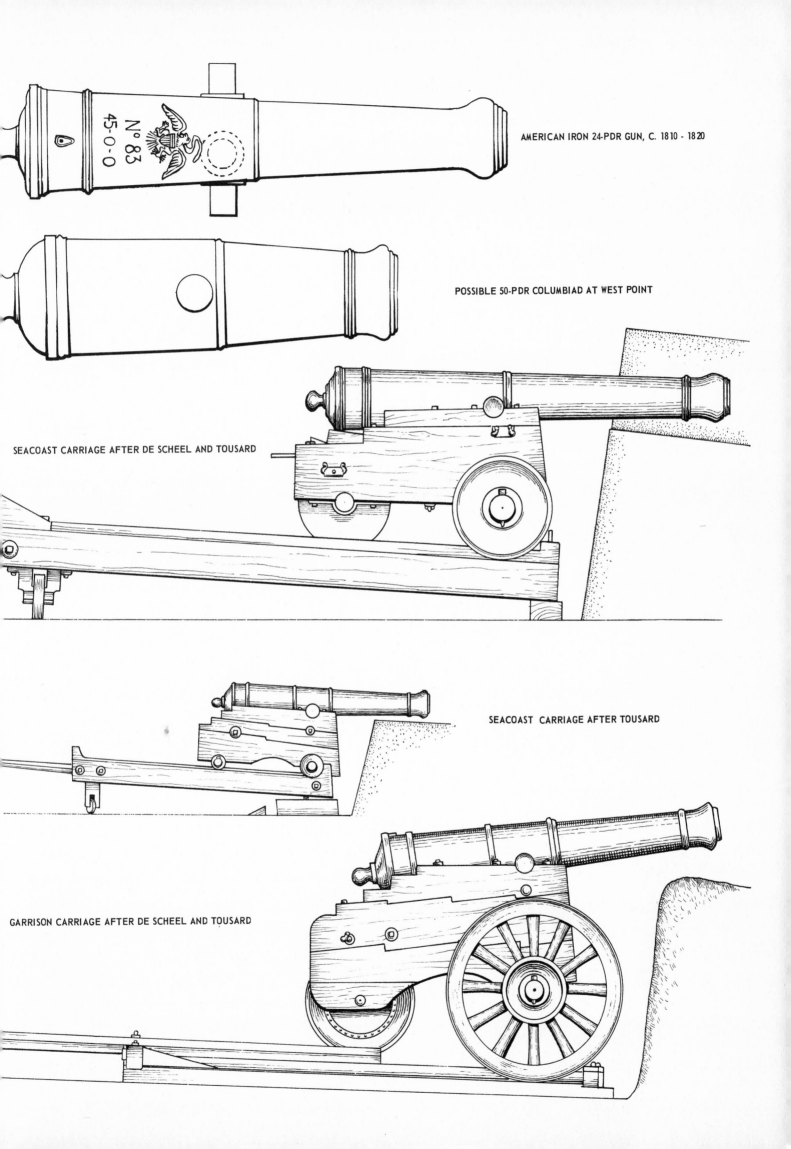

AMERICAN IRON 24-PDR GUN, C. 1810 - 1820

No 83
45-0-0

POSSIBLE 50-PDR COLUMBIAD AT WEST POINT

SEACOAST CARRIAGE AFTER DE SCHEEL AND TOUSARD

SEACOAST CARRIAGE AFTER TOUSARD

GARRISON CARRIAGE AFTER DE SCHEEL AND TOUSARD

it wanted. Some French brass pieces and carriages of the Gribeauval system came to the United States with the Louisiana Purchase, and a few of these participated in the Battle of New Orleans, but generally this was the iron age of cannon in America.

The French Influence on Carriages

Although British practices dominated the field of tube design, American carriages shifted to the French Gribeauval almost completely. Beginning gradually about 1800 one French feature after another made its appearance. The wrought iron axle replaced the wooden version. The penthouse ammunition box between the trails superseded the old side boxes, and the shape of the trails themselves began to follow the French. By 1809 the transition was complete and American carriages were entirely of the Gribeauval pattern with one exception. The tire on the wheel was almost always made in one piece and shrunk in place in this country, and the Gribeauval system of separate segments of tire never became popular.

American artillerymen generally painted these carriages blue as they had during the late years of the Revolution. Repeated orders for Prussian blue and white lead indicate that this was the common mixture for paint. But there is one exception of unknown significance. In 1805 Henry Dearborn instructed one of his ordnance officials to paint all the carriages under his supervision a bright red! There is also one reference to painting a series of carriages blue with red wheels, but these were special guns destined as tribute to the Dey of Algiers. Perhaps no other American carriages were so painted, but they must have been a handsome lot indeed when the Dey received them. Whether blue or red or a combination, American carriages were colorful in these early years, and they mounted their pieces just as well as the drab versions that succeeded in the next period. They were the last of a long tradition of brightly painted mounts that had brought color to artillery since the very beginning.

Seacoast Artillery

AMERICAN ARTILLERY of the federal period breaks down largely into two classes: field and seacoast. The other familiar category of siege and garrison pieces fell into disuse. Theoretically the 18-, 24-, and even the 32-pounders could have been mounted on siege carriages of the Gribeauval type and used for siege work, but the United States conducted no sieges in these years, and so probably no siege carriages were ever constructed for other than test or demonstration purposes. Furthermore most of the western forts employed field guns. They were mainly small works expecting only Indian attack, and the lighter pieces fitted their needs better than the big guns. Thus heavy artillery was confined largely to the fortifications along the seacoast or left in various ordnance yards waiting service if called for.

For this seacoast work the United States relied upon the three biggest calibers it recognized in the beginning—the 18's, 24's, and 32's. As in the field artillery some of these guns were left over from the Revolution or even before, but others were cast at least as early as the 1790's. These American-made big guns were the first in the new nation to be made of iron, and they continued to be iron pieces throughout the period. In fact, no seacoast gun was ever made of brass in the United States afterwards.

In design these big pieces generally followed the same pattern as the field guns. British designs set the standard at first with a modified shift towards French practices after 1800, and then, about 1810 or 1812, came the smooth type without second reinforce, breech ring, or chase girdle, very much like the French naval guns shown by Louis Tousard in his *American Artillerist's Companion* of 1809. Most of these were plain and undecorated pieces, but two 24-pounders survive in Savannah with handsome American eagles in relief just ahead of the vents.

Some actual French guns also found their way

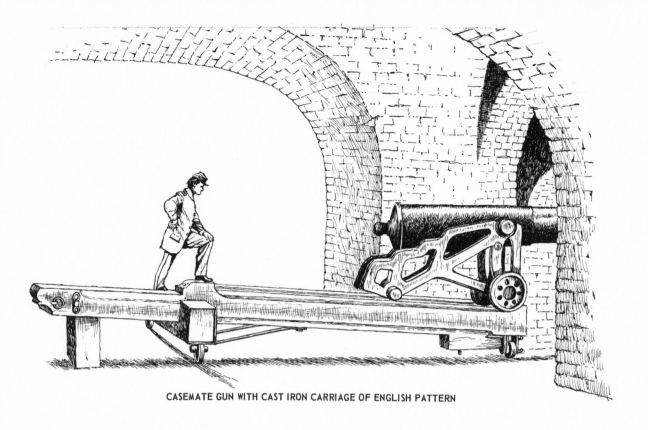

CASEMATE GUN WITH CAST IRON CARRIAGE OF ENGLISH PATTERN

GUN ON WOODEN CASEMATE CARRIAGE, C. 1830

GUN ON CAST IRON CASEMATE CARRIAGE AS DRAWN AT FORT ADAMS, 1835

CARRONADE MOUNTED AS FLANK DEFENCE GUN, DRAWN AT FORT ADAMS, 1835

into American forts. A few of them were mounted at Fort McHenry and replied to the British bombardment in that famous engagement of 1814. Most of these French guns seem to have come from ship armaments, and so they were probably of the French naval pattern. Captured English guns were undoubtedly utilized as well, and they were handier to have since they were built in the same calibers as America's. As noted before, French measurements were slightly larger than English and American, and so there was no interchangeability of ammunition. The occasional references to odd sizes such as 36-pounders in contemporary records quite probably indicate the use of a French gun, though there is some indication that American cannon in the 36-pounder size were contemplated, probably to round out batteries that already had such French pieces. There are also ordnance lists showing 42-pounders and 50-pounders, but here there is some ambiguity because the armament at the same fort is noted one year as 42's and the next year as 50's. Since the guns obviously had not changed in the interval, it would seem that the men who made the measurements differed.

The Riddle of the Columbiad

These size differences and the lack of standardization present the student of artillery with several problems when he tries to analyze the complete system of U. S. artillery during the period. But this is nothing when compared with the other problem of American heavy artillery of the early 1800's—what was the columbiad? This ambiguous term has been used by writers on artillery for almost 150 years, often with little or no understanding of what they were talking about. The result has been almost impenetrable confusion. Most dictionaries and technical works state that a columbiad was a shell gun invented by George Bomford and named after Joel Barlow's patriotic epic *The Columbiad,* which was published in 1809. Actually the term seems to have meant different guns at different times, and it may not have derived from the poem after all.

Almost every serious artillery historian of the early federal period has had a go at untangling the web, and even this brief introduction shall prove no exception. The first known use of the name "columbiad" appears in 1809, and it refers to long guns—some of them 18-pounders, some as small as 6-pounders. Thereafter the word seems to

be applied to a number of different caliber guns, the only common aspect of which is that all were made in America. This suggests that perhaps the name came first from the big guns being cast at the Columbia foundry in Georgetown, D.C., which supplied a big part of the American heavy artillery, and from thence it spread to all American-made guns, at least for a short period. Then, about 1811, 50-pounder guns began to appear, and it is possible that some of these were shell guns, and they may have been invented by Bomford. The practice of firing shells from guns with a flat trajectory had been tested by the British at Gibraltar in 1779, but it was still highly experimental and did not become standard in Europe until the Paixhans gun at nearly mid-century. If Bomford did, indeed, develop a gun for firing shells, it could well be the invention that is mentioned by writers of the middle and late 1800's, and this would tie in with the later columbiads after 1840, all of which were essentially shell guns. Epaphras Hoyt, writing as early as 1811, states that this was the case, and adds also that they were short guns cast at the Columbia Foundry. To support this theory further, one short shell gun or huge howitzer in the 50-pounder size exists at West Point and it seems to date from this period. For perhaps ten years 50-pounder columbiads are listed in some quantity at various forts and yards, and there are even a very few 100-pounders. Then, about 1820, the term drops almost completely out of use and does not reappear until about 1840 at a time, significantly, when George Bomford was Chief of Ordnance.

Thus, to summarize, the puzzling columbiad seems to have been any American gun at first and then later specifically a short shell gun of large caliber used for seacoast defence, probably invented by George Bomford, who resurrected it again when he became Chief of Ordnance. But there is no definite proof.

Barbette and Casemate Carriages

The carriages on which seacoast guns were mounted also present some perplexing problems. The French Gribeauval system had developed a series of carriages for fortifications with pivoted chassis for both barbette and casemate emplacements, and it is probable that Americans adopted these models. They are illustrated in both De Scheel and Tousard, the two available American manuals on artillery for the period, and one little

3'

2'

1'

0'

sketch on a map of Fort Lafayette made in 1815 seems to show a casemate carriage of the type Tousard pictures. There are also references to guns being mounted on barbette carriages at Fort McHenry, and this use of the French term might indicate the presence of Gribeauval carriages. One factor that further complicates the picture of American seacoast carriages is that each one had to be built at the site it was to occupy. Forts were not built with standard parapets or casemates, and so the carriage had to be constructed to fit in each instance. This naturally led to some variations and prevented any complete standardization.

These carriages were a great improvement over the older patterns which employed simple naval-type truck carriages. It had always been a problem to traverse these older carriages. The wheels were designed for recoil only and they had to be slid along the floor sideways with pressure from handspikes or tugs from rope rigging to change aim laterally. With the new pivoted chassis the gun could be traversed rapidly and easily, while an upper carriage sliding along tracks on top of the chassis handled all the problems of recoil and return to battery.

Early carriages of the French pattern were made of wood, sometimes with iron tracks for the recoil. As years went by there were various modifications that are not specified in official documents, but there is a drawing in the National Archives made apparently in the 1830's which shows a highly developed casemate carriage quite similar to those used during the Civil War. Meantime the Chief of Ordnance, Decius Wadsworth, had issued an order in 1818 for seacoast carriages of cast iron. Supposedly all new carriages were to be of this material, and the documents indicate that many were actually made. Unfortunately there is no description, and no specimens survive, but there are two bits of evidence that help deduce what at least the casemate carriages looked like. First, there is a drawing made at Fort Adams, Rhode Island in 1835, which illustrates a very simple iron frame on top of a wooden chassis, apparently a purely American design. At the same time, however, guns on iron carriages were also installed at Fort Pulaski near Savannah, Georgia, and a photograph of one of the ruined casemates after the Union bombardment in 1862 shows an iron carriage of British design with higher cheeks and more complicated bracing. Thus it would seem that both types were in use simultaneously, and perhaps there were other designs as well.

Carronades on Land

One final and intriguing development in seacoast armament was the use of naval carronades for flank defence. Carronades were short guns of heavy bore designed originally for battering purposes. When placed in casemates commanding the drawbridge and main entrance to a fort, however, they would have been used for antipersonnel work rather than battering. Thus their large bores would have been loaded with grape or cannister. Howitzers were customarily used for this purpose, but there is no reason why the carronade would not have been just as good, and apparently American artillerymen happily adopted the naval piece for their use.

This was the story of American heavy artillery in the years from the close of the Revolution to 1835. It is a confused picture with many questions that are still unanswered, for it was an era of growth and experimentation that no contemporary observer bothered to record in detail. Yet it is an important period, for these years saw the growth of America's seacoast defences from isolated posts to an integrated system of defence that became perhaps the strongest in the world. The cannon were an integral part of that system, and they deserve the continued attention of scholars who may someday solve at least some of the problems mentioned here.

Ammunition—Old Traditions and New Departures

IN THE MAIN ammunition followed exactly the same designs during the federal period that it had during the American Revolution. Solid shot, simple shell, grape, and cannister remained the principal projectiles with perhaps a few carcasses. Powder charges were

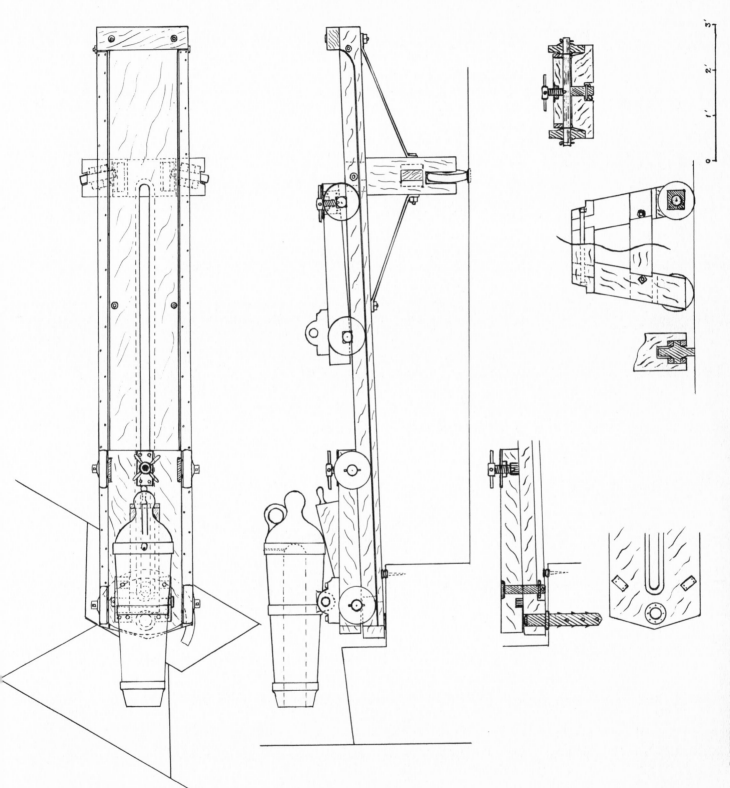

PATTERN FOR 24-PDR CARRONADE MOUNTED AS FLANK DEFENCE GUN, C. 1835

0   1'   2'   3'

wrapped in flannel cartridges, and for rapid fire in field guns these cartridges were fastened to the sabots of the projectiles to make fixed rounds. For the most part there was literally nothing new.

Spherical Case

But there was one very important innovation on the horizon, and it is just possible that it may have been adopted by the American field artillery in this period. This was spherical case or shrapnel. Back in 1784 Lieutenant Henry Shrapnel of the Royal Artillery in England had developed a new projectile. It consisted of a simple shell filled with musket balls and holding a small bursting charge in the center. This seems simple enough, but it offered tremendous possibilities for antipersonnel work. The usual charges for use against enemy infantry were grape and cannister, and occasionally shell. The first two types were remarkably effective at close range, but beyond 500 yards their individual shot had spread out so widely that they did little damage. Also these small shot lost momentum rapidly. Shell carried much farther, but it fragmented into only a few pieces and so did limited damage. Shrapnel's invention offered the advantages of cannister charges at long range. Theoretically the shell carried the shot to a position just over and in front of the target, then burst and released its deadly hail of musket balls which spread out in a cone of destruction—and this is just how it did work when the cannoneers set the fuze properly.

As the years passed there were gradual improvements. The walls of the casing were made thinner than on regular shell, and the musket balls were imbedded in a mass of melted resin or sulphur to prevent them from rattling around and perhaps cracking the shell prematurely. Also the fuze hole was narrowed down at its inner end by a thin plate to concentrate the force of the rupturing charge. This charge also was reduced so that it was just sufficient to break the case and not so strong that it fanned the balls out too widely.

It took a few years for the British Ordnance to become interested in Shrapnel's invention, but by about 1804 or 1805 they had realized its usefulness and its practicality and brought it into play in the Peninsular Campaign against the French. No one knows when the new projectile came to America, but certainly American officers became aware of it soon after its employment in Spain. The *Ordnance Manual* of 1841 lists it as a standard item, so it was definitely in use by then, and the best estimate is that it had been tested and employed in American field guns a good bit earlier. If so, it brought at least one innovation in the ammunition field to the federal period.

The New Gun Drill

IF ANYTHING there was even less change in artillery implements and instruments than there had been in ammunition. The actual shapes of a few implements varied with new or different decorative turnings on portfire stocks and linstocks, but there were no new tools or instruments. Guns and mortars were still loaded, aimed, and fired by the same means.

There was, however, some change in loading technique. French drill seems to have replaced the English routine in at least some instances. Most notably with field pieces this involved shifting the cannoneer (or matross as he was usually called) who held the portfire and actually touched off the piece from the left to the right side of the breech. This meant that he had to stand with his back to the muzzle in order to keep his right hand with the portfire towards the gun. At the same time the matross who thumbed the vent moved to the left of the breech, and in some drills he did the actual thumbing with his left hand, using the right hand to adjust the elevating screw at the direction of the gunner.

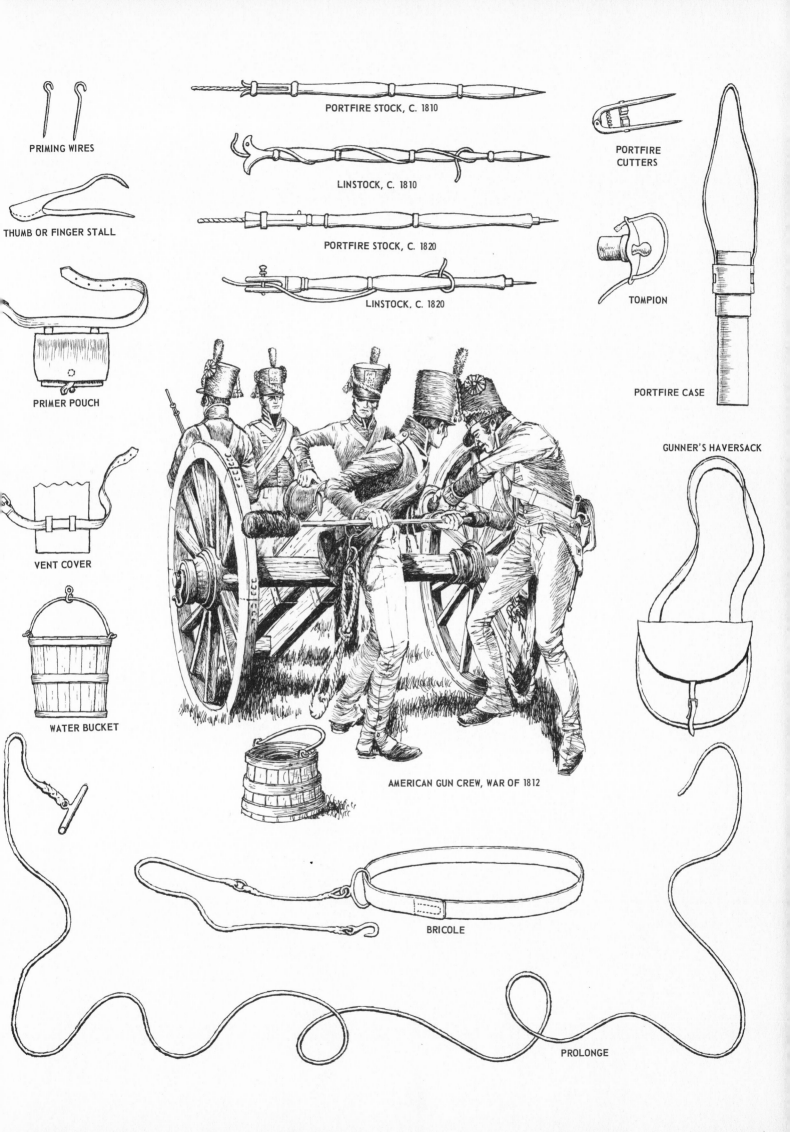

PRIMING WIRES

THUMB OR FINGER STALL

PRIMER POUCH

VENT COVER

WATER BUCKET

PORTFIRE STOCK, C. 1810

LINSTOCK, C. 1810

PORTFIRE STOCK, C. 1820

LINSTOCK, C. 1820

PORTFIRE CUTTERS

TOMPION

PORTFIRE CASE

GUNNER'S HAVERSACK

AMERICAN GUN CREW, WAR OF 1812

BRICOLE

PROLONGE

These were the techniques used by American artillerymen who manned their pieces from St. Louis and New Orleans in the new Louisiana Territory to the coast of Maine and all along the Atlantic seaboard, even to Florida, where the ancient Spanish Castillo de San Marcos had come into United States possession in 1821 and been renamed Fort Marion. Regular or militia, they changed from the old English system to the newer French idea for both materiel and drill. Then change loomed again. Even as U. S. artillerists examined the old Spanish guns at Fort Marion, experiments began with new carriages and new guns. The first of a new series of artillery boards was appointed in 1831, and the studies it began were carried further by succeeding boards until the whole American system of artillery stood on the threshold of a new era in 1835.

CHAPTER IV

The Apex of the Muzzle-Loader, 1836-1865

In the thirty years after 1836 muzzle-loading artillery in America reached its highest peak of development. Newly designed tubes offered ever stronger and lighter pieces. The stock trail carriage brought greater mobility and maneuverability to field artillery, and a whole host of imaginative and ingenious new projectiles made rifled guns practical for the first time. With the rifles came greater range, accuracy, and penetrating power that brought a new dimension to warfare. Yet smoothbores still remained important, and in some situations they even outperformed the sleek new rifles. This was especially true of close-range antipersonnel work, but it also held for some counterbattery fire as well. Thus the smoothbores and rifles stood hub to hub throughout the period.

The new tubes, the rifles, the stock trail carriage, and improved projectiles brought the muzzle-loaders to the zenith. To many a cannoneer the cannon of the Civil War must have seemed perfection indeed. But even then the seeds of change were germinating. Radically different cannon that loaded at the breech appeared and saw limited service in the field. These were largely experimental pieces. Many flaws remained to be smoothed out, but they pointed the way to the future even though they did not displace the muzzle-loader immediately. The years after 1865 saw the older patterns fade in importance as more and more of the new breechloaders came into use. A few of the old muzzle-loaders mustered for the Spanish-American War, and some still served for National

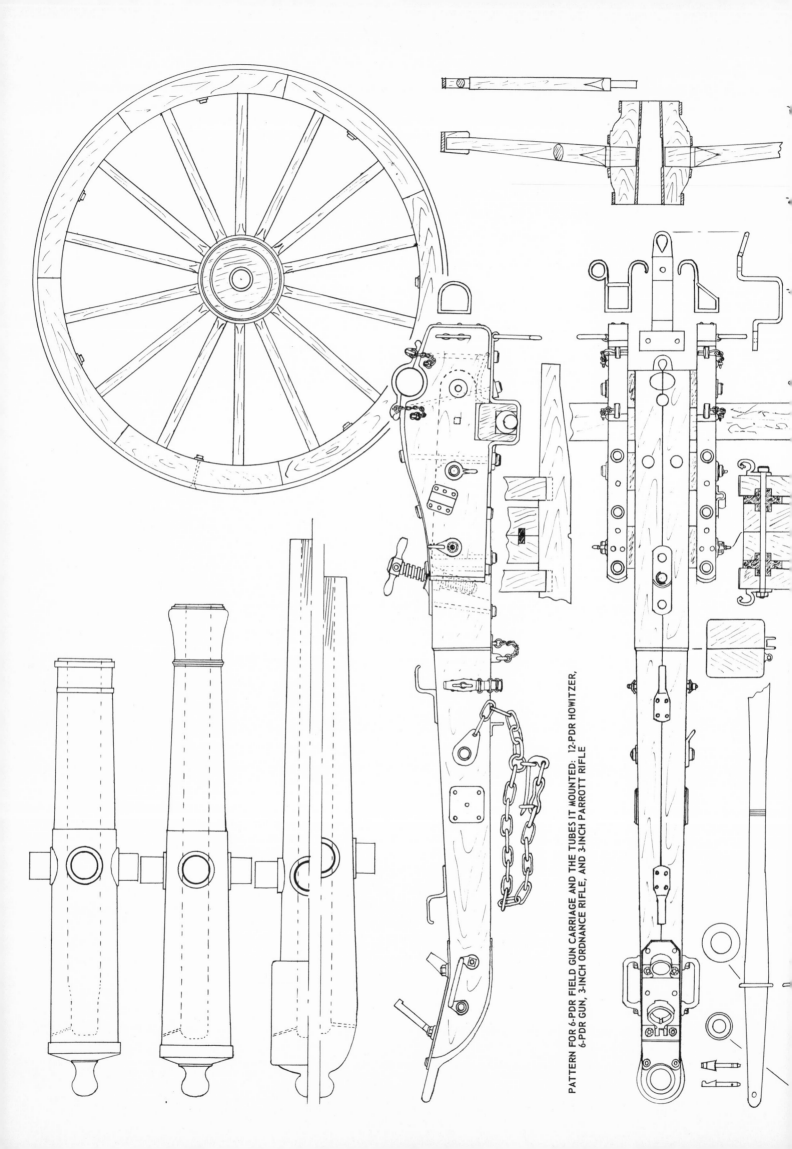

PATTERN FOR 6-PDR FIELD GUN CARRIAGE AND THE TUBES IT MOUNTED: 12-PDR HOWITZER,
6-PDR GUN, 3-INCH ORDNANCE RIFLE, AND 3-INCH PARROTT RIFLE

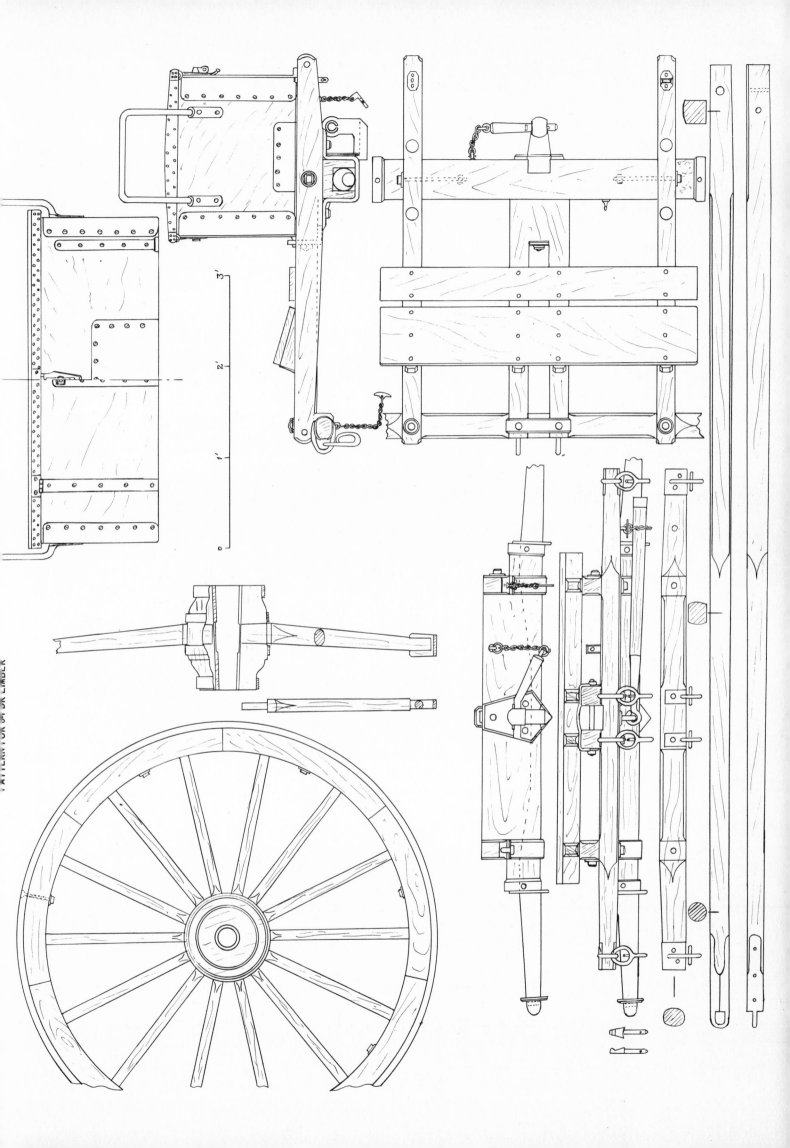

3'

2'

1'

Guard training as late as 1907. These were mere relics, however. None had been made since 1865. The Civil War marked the real end of their era, and these veterans soon joined their comrades in battlefield parks or on courthouse lawns, silent reminders of a glorious past.

A Glorious Era

AND A glorious era it was indeed. American artillerymen, who had been so often frustrated and left gunless in the early years of the century, came into their own at last. In the Mexican War and again in the Civil War they built a legend of daring that added name after name to the pantheon of American heroes. In the nineteen years between 1846 and 1865 American artillerymen set a standard for gallantry that has never been surpassed.

Gunnery Heroes of the Mexican War

It all started with the battle of Palo Alto, May 8, 1846. An outnumbered American force under General Zachary Taylor faced a determined Mexican foe in Texas. Repeated infantry and cavalry attacks by the enemy threatened Taylor's small army time and again, but American artillery proved decisive. Major Samuel Ringgold with his famous "flying battery" pounded the foe with his bronze 6-pounders, and at one point Captain James Duncan daringly advanced with his battery and enfiladed the whole Mexican line. No force could withstand the concentrated shot, and the Mexicans fell back. But Major Ringgold fell, too, shot through both thighs by a Mexican cannon ball, and with his tragic death the nation gained a hero, giving his name to numerous streets and towns throughout the land.

The next year at Buena Vista American artillery did it again. Braxton Bragg's battery loaded with double cannister and smashed a surging Mexican attack, helping to wring victory from what looked like sure defeat. And there were others—John Paul Jones O'Brien who lost his 6-pounder guns with honor at Buena Vista, Thomas J. Jackson, not yet called "Stonewall," who cleared the way at Chapultepec, "Prince John" Magruder, to name but a few.

Gunnery Heroes of the Civil War

In the Civil War there was another host of heroes. Improved rifle muskets had made the cannoneer's life almost worthless at less than a thousand yards. Yet time after time batteries unlimbered in the very face of the deadly minié balls. At Antietam Stewart's battery manned its guns less than 50 yards from the Confederate lines, and the cannister from their Napoleons harvested the storied cornfield, driving the Rebels back despite their grim determination. They lost 40 men out of 100 in just a few minutes, but they could still cheer when another old artilleryman, John Gibbon, forgot his general's stars and helped them man one of their stricken pieces.

Both sides produced a bumper crop of gallant gunners in that bloody war. At Harpers Ferry near-sighted "Willie" Pegram moved his gun far to the front—just so he could see the blue-clad enemy, they said. It became a habit. At Fredericksburg in 1862 young John Pelham's two Napoleons darted about in the mist and held up the advance of three Union divisions as a host of Yankee cannon sought to take him in counterbattery. The next year at Chancellorsville Captain Hubert "Leatherbreeches" Dilger delayed the advance of Jackson's famed "Foot Cavalry" long enough for the Union retreat to be made good. Bouncing his shot down the turnpike at the advancing troops, he doggedly retreated "by the recoil of his pieces." He lost one of his guns, but he helped save the day. At Gettysburg there were more names—Cushing, Hazlett, Stewart, Dilger again. Most of them Union men, for artillery is most gallant in defence. Attack support and counter-

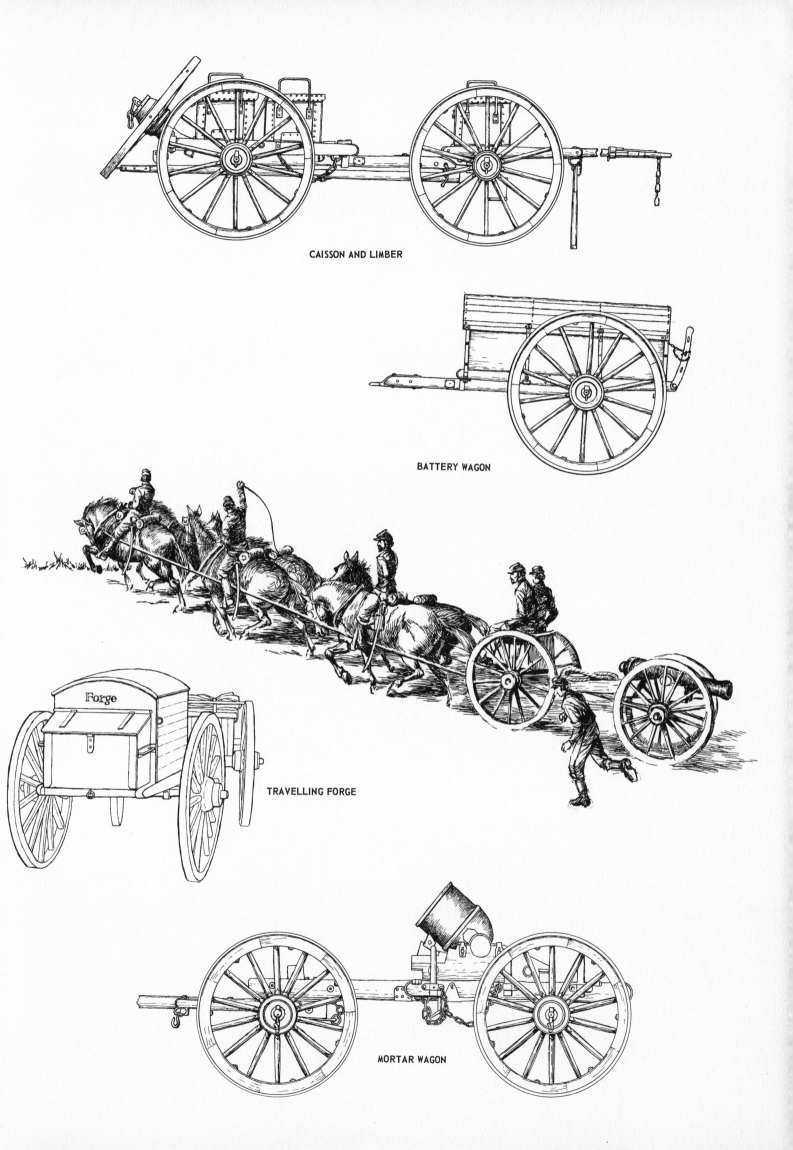

CAISSON AND LIMBER

BATTERY WAGON

TRAVELLING FORGE

MORTAR WAGON

battery are important, but it is the battery that stands fast in the face of an infantry charge that attracts notice. It also attracts casualties, and both Hazlett and Cushing fell—Cushing with five wounds, literally "holding his guts in his hands" as he got off his last shot.

Out in the West, cannon played their part, too. There it was the little mountain howitzer that accompanied expeditions and held forts. And it won battles. Apache Pass was decided by the "wagon guns" in 1862 and so was First Adobe Walls in 1864. Kit Carson welcomed the little weapons, and so did the Regular officers. The Indians feared them.

In siege artillery the new rifles changed the face of warfare. At Fort Pulaski in 1862, huge Union rifles quickly shattered the brick walls that Confederates had thought almost impregnable. In a few short hours the big guns proved that masonry forts were a thing of the past. Rifles could destroy them from ranges the engineers who built them thought impossible.

Such was the role of the finest American muzzle-loading artillery. It played a stellar role in American expansion, in the winning of the West, and in the tragic years of fraternal strife. The guns were good; the men who fought them were superb. It made an impressive combination, and it left an indelible imprint on American history.

Changes in Field Artillery

FIELD ARTILLERY witnessed three developments of profound importance in the years from 1836 to 1865. First there was the introduction of the stock trail carriage, second came the return to bronze as the metal for field cannon, and finally practical rifles made their appearance. None was a sudden event bursting full-blown upon the scene. Each represented the end product of years of tests and thought.

The Stock Trail Carriage

The stock trail development, in fact, began back before 1800. As early as 1778 British artillery designers had begun to experiment with carriages boasting a single sturdy trail in place of the split or flask trail design then popular in all the major artillery systems of the world. The new design offered several advantages. It was at least as strong if not stronger than the split trail. It was simpler to make, and it afforded a shorter turning radius for the carriage, thus offering greater maneuverability. Americans encountered carriages of the new English design during the War of 1812, and actually captured some of them at Plattsburg. Yet few American artillerymen seemed impressed with the advantages of the new design. They had, after all, just adopted the Gribeauval pattern carriage and were thoroughly imbued with the genius of that great artillerist. There was one exception to this general attitude, however. Decius Wadsworth, first Chief of Ordnance, liked the English carriages. He ordered several modified versions of them manufactured for tests, and he recommended their adoption. The consensus was against him, unfortunately, and the artillery board of 1818 rejected the idea.

The board's decision was a delay rather than a defeat. The stock trail had too many advantages to be denied completely. The French began to copy the English design and officially adopted a modified version of it in 1827. In 1829 Lt. Daniel Tyler provided drawings of the French carriages to the United States, and in 1830 work began on experimental copies of them. These proved so successful that Secretary of War Lewis Cass offi-

TABLE OF FIRE. 6 POUNDER GUN

SHOT
Charge 1¼ Pounds

ELEVATION In Degrees	RANGE In Yards
0°	318
1°	674
2°	867
3°	1138
4°	1256
5°	1523

SPHERICAL CASE
Charge 1¼ Pounds

ELEVATION In Degrees	TIME OF FLIGHT In Seconds	RANGE In Yards
1° 0′	2″	600
1° 45′	2″75	700
2° 0′	3″	800
2° 45′	3″25	900
3° 0′	3″75	1000
3° 15′	4″	1100
4° 0′	5″	1200

Use SHOT at masses of troops, and to batter, from 600 up to 2.000 yards. Use SHELL for firing buildings, at troops posted in woods, in pursuit, and to produce a moral rather than a physical effect; greatest effective range 1,500 yards. Use SPHERICAL CASE SHOT at masses of troops, at not less than 500 yards; generally up to 1,500 yards. CANISTER is not effective at 600 yards; it should not be used beyond 500 yards, and but very seldom and over the most favorable ground at that distance; at short ranges, (less than 200 yards,) in emergency, use double canister, with single charge. Do not employ RICOCHET at less distance than 1,000 to 1,100 yards.

CARE OF AMMUNITION CHEST.

1st. Keep everything out that does not belong in them, except a bunch of cord or wire for breakage; beware of loose tacks, nails, bolts, c'' scraps.
2d. Keep friction primers in their papers, tied up. The pouch containing those for instant service must be closed, and so placed as to be secure. Take every precaution that primers do not get loose; a single one may cause an explosion. Use plenty of tow in packing.

(This sheet is to be glued on to the inside of Limber Chest Cover.)

TABLE OF FIRE. LIGHT 12-POUNDER GUN. MODEL 1857.

SHOT. Charge 2¼ Pounds.		SPHERICAL CASE SHOT. Charge 2½ Pounds.			SHELL. Charge 2 Pounds.		
ELEVATION In Degrees.	RANGE In Yards.	ELEVATION In Degrees.	TIME OF FLIGHT. Seconds.	RANGE In Yards.	ELEVATION In Degrees.	TIME OF FLIGHT. In Seconds.	RANGE In Yards.
0°	323	0°50'	1"	300	0°	0"75	300
1°	620	1°	1"75	575	0°30'	1"25	425
2°	875	1°30'	2"5	635	1°	1"75	615
3°	1200	2°	3"	730	1°30'	2"25	700
4°	1325	3°	4"	960	2°	2"75	785
5°	1680	3°30'	4"75	1080	2°30'	3"5	925
		3°40'	5"	1135	3°	4"	1080
					3°45'	5"	1300

Use SHOT at masses of troops, and to batter, from 600 up to 2.000 yards. Use SHELL for firing buildings, at troops posted in woods, in pursuit, and to produce a moral rather than a physical effect; greatest effective range 1,500 yards. Use SPHERICAL CASE SHOT at masses of troops, at not less than 500 yards; generally up to 1,500 yards. CANISTER is not effective at 600 yards; it should not be used beyond 500 yards, and but very seldom and over the most favorable ground at that distance ; at short ranges, (less than 200 yards,) in emergency, use double canister, with single charge. Do not employ RICOCHET at less distance than 1,000 to 1,100 yards.

CARE OF AMMUNITION CHEST.

1st. Keep everything out that does not belong in them, except a bunch of cord or wire for breakage ; beware of loose tacks, nails, bolts, or scraps.
2d. Keep friction primers in their papers, tied up. The pouch containing those for instant service must be closed, and so placed as to be secure.
Take every precaution that primers do not get loose ; a single one may cause an explosion. Use plenty of tow in packing.

(This sheet is to be glued on to the inside of Limber Chest Cover.)

cially approved them for United States service in 1836. Full production, however, was delayed, and the standard stock trail carriage when it appeared was officially designated the pattern of 1840.

Generally these stock trail field carriages came in three sizes. The most popular one, called the 6-pounder gun carriage, mounted the 6-pounder itself, the 12-pounder howitzer, and later the 3-inch Parrott and Ordnance rifles plus some of the Blakely and Wiard rifles. The next larger size was known as the 24-pounder howitzer carriage, but it also mounted the Napoleon. Quite apart was the carriage or series of carriages designed for the mountain howitzer and described below. And then there were a few special carriages, one designed but built only in small numbers for the Wiard, a Whitworth carriage, and an Armstrong carriage. There was also a carriage for the big 12-pounder gun and the 32-pounder howitzer, but these appeared so seldom that they can practically be disregarded.

Closely allied to the carriages was the rest of the rolling stock. A new limber holding the ammunition chest was adopted, and there were fixed iron handles on the chest so designed that cannoneers could ride on it and hold on as the springless vehicle bounced along behind a six-horse team. Two more ammunition chests and a spare wheel were carried on the caisson, which was also a two-wheeled vehicle and required a limber of its own for driving. In action the limber was stationed to the rear of the gun's trail with the tongue pointing towards the cannon. The caissons were parked still further to the rear. In addition there was a new travelling forge for field repairs and horseshoeing, a battery wagon for extra supplies and equipment, and various additional vehicles such as mortar wagons for transporting heavy pieces. In action these stayed well to the rear. Only the limbers and some of the caissons moved to the front.

Conversion from Iron to Bronze

The adoption of bronze as the metal for field pieces followed a rocky course similar to that of the stock trail carriage. The iron "walking stick" models of 1818 had proved undependable, and some of them had burst during tests in 1827, causing practical artillerymen to lose faith in them and in iron in general as a proper metal for field guns. It was, they believed, too heavy and too brittle. The subject was high on the agenda of

the board of ordnance appointed by Secretary Cass in 1831, and it ordered a full study of the problem which continued until 1835, when a committee of highly talented officers flatly recommended that iron be rejected in favor of bronze. Cass approved the recommendation, but the iron founders of the nation fought hard against the decision. They were reinforced by the fact that iron was still the only material readily available from native sources, and Secretary of War Poinsett, who succeeded Cass, was reluctant to make the United States dependent upon imports for such a strategic material. Nevertheless some bronze guns were produced in 1836, and he finally approved bronze as standard in 1841.

Mordecai's New System of Artillery

With the acceptance of bronze the United States also adopted a whole new system of field artillery materiel. Many experienced officers worked on the problem, but most of the detailed work fell to Alfred Mordecai. It was he who determined the final proportions of tubes and carriages, and he oversaw the completion of a whole volume of text plus an atlas of plates setting forth every detail of the construction of tubes, carriages, implements, instruments, and ammunition, even harness for the horses, tying it down to the final rivet. It was a mammoth undertaking, but it gave the country its first true system of artillery materiel.

For field artillery, this system included 6- and 12-pounder guns, 12-, 24-, and 32-pounder howitzers, and a 12-pounder mountain howitzer. Of these pieces the 12-pounder gun and the 32-pounder howitzer were little used. They were too big and awkward. The 6-pounder gun was the principal weapon for most field batteries. It was the prime artillery weapon for the Mexican War and saw wide service during the opening years of the Civil War, especially on the Confederate side. Then it was superseded at least in the Army of the Potomac by the light 12-pounder gun, model 1857.

The Versatile Napoleon

The Napoleon, as the model 1857 was generally called, was a truly fine weapon. Artillerists had long felt the need for a gun somewhat bigger than the 6-pounder but not so heavy as the standard 12-pounder, model 1841. Thus they quickly seized upon a new design sponsored by the

Emperor Louis Napoleon of France. It was designed to replace both guns and howitzers of comparable size in the French establishment, and so it was frequently called a gun-howitzer although it had no howitzer characteristics whatsoever. It could fire cannister and shell as well as the howitzer, and its range with solid shot was just as great as the old 12-pounder gun. Yet the tube weighed about 1,200 pounds, 500 pounds less than the model 1841. It was also a shorter weapon and infinitely more mobile.

Still it took some time for the Napoleon to catch on. The first gun of the type was cast early in 1857. It proved to have some drawbacks, and so it was redesigned very slightly. Four more were then cast, and these five Napoleons remained the only ones in the United States service until 1861. Then production began in earnest. The dolphins or handles which had graced the early models were ordered omitted late in 1861. By that time only about 36 of the guns had been cast, but by the end of the war 1156 Napoleons had appeared in the North and perhaps another 630 in the Confederacy. At the battle of Gettysburg all but one of the batteries in the Army of the Potomac were armed either with Napoleons or rifles. All other types of artillery had disappeared. The preponderance was never so great in the Confederate armies or in the Union armies of the West, where the 6-pounder gun and the 12-pounder howitzer continued in the use to the end of the war.

There were few variations in the Napoleon design after the disappearance of the dolphins. Except for minor variations in contractor's work, the Union specimens remained virtually the same for the entire period. In the Confederacy a type similar to the dolphinless Union model shared popularity with a smoothly tapering version without any muzzle swell. There was also a gun called an iron Napoleon produced by the Tredegar Iron Works in Richmond, but it could be called a Napoleon only because it was intended for the same purposes. It was banded at the breech and lcoked somewhat similar to the standard Parrott rifle of the period. In the North at least six Napoleons were made as rifled pieces on an experimental basis, but they never became standard, and their service history is completely unknown.

Rifled Artillery

The rifled Napoleon represented an attempt to obtain for that already versatile weapon the further advantages of the rifle's accuracy. For many years artillerists had coveted the accuracy of the rifled small arm, but they had been hampered because no projectile had been designed that could be loaded easily yet fit the bore tightly enough in firing to take the spin imparted by the spiral grooves of the rifling. Finally, just about at mid-century, a number of projectile designers suddenly succeeded almost at the same time. Their efforts will be discussed in the section on ammunition, but their work did result in new rifled tubes being added to the United States artillery system.

One of the designers who worked on rifle projectiles was General Charles T. James, who also devised a method for converting existing bronze smoothbores to rifled cannon. His system involved a number of narrow, deep grooves that could be cut into the bores of these older pieces, which then became known as James rifles. Their size designation changed with their name. Because the elongated rifle projectiles weighed about twice as much as the spherical balls they had formerly thrown, a rifled 6-pounder became known as a 12-pounder James, and a 12-pounder became a 24-pounder. James also produced a few bronze rifles that had never been smoothbores. These were sleek pieces with smooth lines somewhat like the contemporary iron Ordnance rifles of the period. The James system seems to have worked relatively well, but it had at least two drawbacks. Since the grooves were narrow and deep, smoldering fragments of cartridge might possibly go unquenched in their depths and cause a premature discharge when the next charge was loaded, and since bronze was a soft metal, the James rifles tended to wear out in use fairly quickly.

Iron was a much more durable material, and so iron rifles quickly came to dominate the field. Two patterns predominated in the federal service, the Parrott and the Ordnance rifle. Robert P. Parrott developed his design in the 1850's, and obtained a patent on it in 1861. It consisted of a cast iron tube with a wrought iron band shrunk around the breech at the point of greatest pressure from the charge within. These Parrott rifles were cast at the West Point Foundry, which Parrott supervised, and for field use were made principally in the 3-inch or 10-pounder size. These were actually about 2.9 inches in bore or a very scant 3 inches. There was also a 20-pounder size with a bore of 3.67 inches, but it was much less used. Both were good guns, but they developed

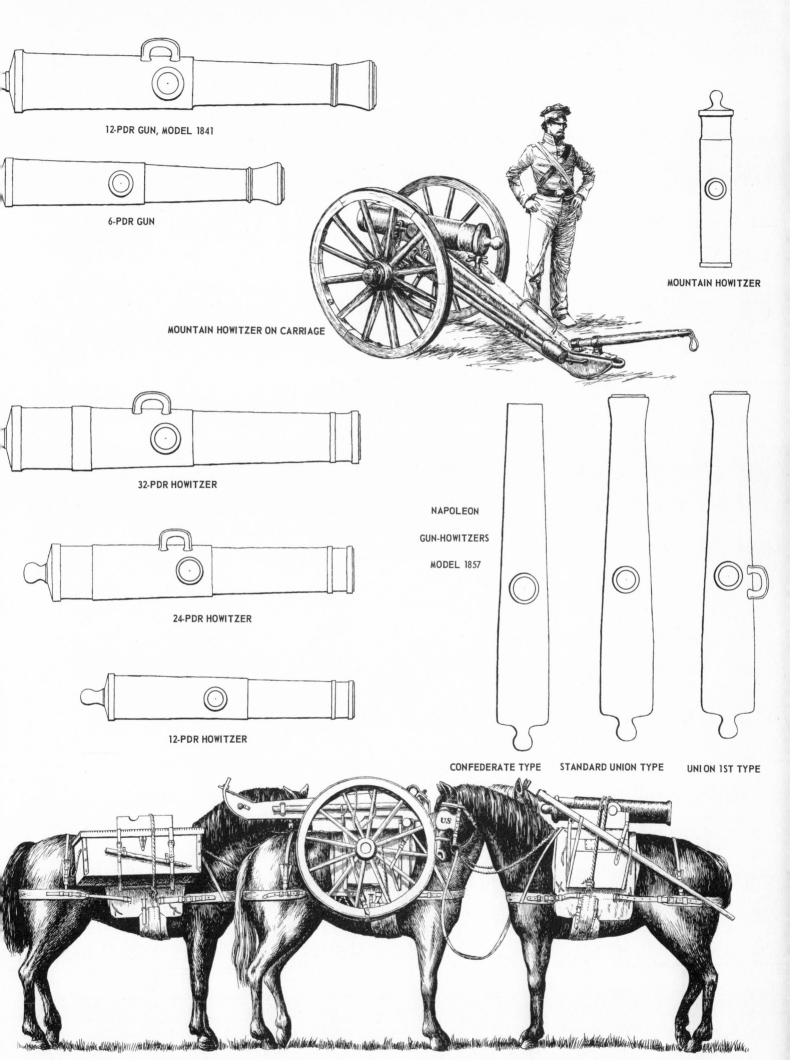

12-PDR GUN, MODEL 1841

6-PDR GUN

MOUNTAIN HOWITZER ON CARRIAGE

MOUNTAIN HOWITZER

32-PDR HOWITZER

24-PDR HOWITZER

12-PDR HOWITZER

NAPOLEON

GUN-HOWITZERS

MODEL 1857

CONFEDERATE TYPE   STANDARD UNION TYPE   UNION 1ST TYPE

MOUNTAIN HOWITZER, CARRIAGE AND AMMUNITION CHESTS PACKED ON THREE HORSES

12-PDR WHITWORTH BREECHLOADING RIFLE

12-PDR WHITWORTH MUZZLE-LOADER

12-PDR BROOKE RIFLE

CONFEDERATE 3-PDR MOUNTAIN RIFLE

WIARD RIFLE ON CARRIAGE

6-PDR WIARD RIFLE

10-PDR WIARD RIFLE

3-INCH ORDNANCE RIFLE

12-PDR ARMSTRONG BREECHLOADING RIFLE

10-PDR ARMSTRONG MUZZLE-LOADING RIFLE

10-PDR PARROTT RIFLE

20-PDR PARROTT RIFLE

10-PDR BLAKELY RIFLE, TYPE I

10-PDR BLAKELY RIFLE, TYPE II

10-PDR JAMES RIFLE

a bad reputation among some artillerymen for bursting just ahead of the breech band after prolonged service.

The 3-inch Ordnance rifle was an even better piece. Made of tougher wrought iron it had no need for a breech reinforcement, and so it tapered evenly from breech to muzzle in a sleek line. For some reason contemporary artillerymen often referred to these pieces as Rodmans although the great Thomas Jefferson Rodman seemingly had nothing to do with either their design or manufacture. Perhaps it was because their sleek lines generally followed the pressure curves that he had worked out for cannon in general and for his big guns in particular. The Ordnance rifle was a full three inches in the bore, and it normally fired Hotchkiss or Schenkl projectiles or even Parrott shells in an emergency, while the Parrott could fire only Parrott or sometimes Schenkl projectiles.

Both the Parrott and the Ordnance rifle were adopted as standard in 1861, and they performed very well. With an elevation of only five degrees they could shoot over 1,800 yards, more than a mile, and with increased elevation they could shoot even further. There was little value to these extreme ranges, however. The idea of indirect firing was just being developed, and it was not understood or used by artillerymen in the field. Thus a gunner could shoot only as far as he could see, and this restricted his practice considerably. For ranges of 1,800-2,000 yards against sizeable targets, however, the rifle's accuracy was extremely useful. For close-range work against charging infantry it could employ cannister like the smoothbores, but its smaller bore made it less effective at close quarters than the Napoleon. Thus a well-armed division or even a brigade would have had some batteries of rifles and some of Napoleons, each for its specific uses.

There were other rifles in service in smaller quantities. Norman Wiard devised 6- and 10-pounder steel rifles with highly advanced carriages. About 44 of these rifles took the field. English Blakely rifles found their way into Confederate batteries, and then there were the spectacular Armstrong and Whitworth rifles. The Armstrongs never became very popular, but a few Whitworths did see service on both sides. These were made both as breechloaders and muzzle-loaders, and they were characterized by hexagonal bores, the sides twisting to give the projectiles their spin. The Whitworths had spectacular ranges and accuracy—better than two miles, but again this was of little practical value because of the state of the science of gunnery. Also their projectiles were so slender that shells could carry only tiny bursting charges, and case shot was almost impossible. Thus they were practical only for battering with solid bolts and so were not nearly as useful as their statistics might indicate at first glance.

The Mountain Howitzer

In addition to the principal rifles and smoothbore guns, there was one other piece in a category all by itself. This was the little 12-pounder mountain howitzer. The need for a light, easily portable piece that could be used in the West had become manifest early in the century. Thus when France developed such a weapon the United States was quick to follow its example, and the first little mountain howitzers appeared in 1836. The tubes of these first pieces were good, but the carriages were too light and proved unserviceable. In 1840 the tube was slightly modified as the model 1841, and a new pack carriage devised. This carriage could be broken down and carried by three horses or mules with the tube on one, the carriage on another, and the ammunition on a third. Shafts could also be attached to the trail so that it could be pulled by one horse. About 1850 a so-called prairie carriage was added with a limber so that it could be pulled more efficiently, and in the 1860's another prairie carriage with a wider axle and thus greater stability appeared.

Some mountain howitzers were used during the Civil War in the East, but they were never popular. They could not compete with the bigger pieces in a country where there were usually passable roads. In the West where there was no counter-battery work to worry about, where portability was of the utmost importance, and where antipersonnel charges against Indians represented the principal use, the little pieces won widespread approval and enthusiasm. They accompanied columns of soldiers in the field, and they armed forts and posts. Some of them continued in use during the Nez Percé Campaign and the Modoc War of the 1870's. A few of these tiny pieces survived even longer, but not so long as some Napoleons and 3-inch Ordnance rifles that can be found still on the active lists for National Guard armament in 1907, the very last of the muzzle-loading field artillery to remain in active service.

FIELD ARTILLERY

SMOOTHBORES	Bore diameter (inches)	Material	Length of tube (inches)	Weight of tube (pounds)	Weight of projectile (pounds)	Weight of charge (pounds)	Muzzle velocity (ft. per sec.)	Range at 5° elevation (yards)
Models of 1841-44								
6-pdr gun	3.67	bronze	60	884	6.10	1.25	1,439	1,523
12-pdr gun	4.62	"	78	1,757	12.30	2.50	1,486	1,663
12-pdr howitzer	4.62	"	53	788	8.90*	1.00	1,054	1,072
24-pdr howitzer	5.82	"	65	1,318	18.40*	2.00	1,060	1,322
32-pdr howitzer	6.40	"	75	1,920	25.60*	2.50	1,100	1,504
12-pdr mtn. howitzer.	4.62	"	32.9	220	8.90*	0.50	650	900
Model of 1857								
12-pdr Napoleon	4.62	"	66	1,227	12.30	2.50	1,440	1,619

*Weight of shell

RIFLES	Bore diameter (inches)	Material	Length of tube (inches)	Weight of tube (pounds)	Weight of projectile (pounds)	Weight of charge (pounds)	Muzzle velocity (ft. per sec.)	Range at 5° elevation (yards)
10-pdr Parrott	3.00	iron	74	890	9.50	1.00	1,230	1,850
3-inch Ordnance .... (Rodman)	3.00	"	69	820	9.50	1.00	1,215	1,830
20-pdr Parrott	3.67	"	84	1,750	20.00	2.00	1,250	1,900
12-pdr James	3.67	bronze	60	875	12.00	0.75	1,000	1,700
24-pdr James	4.62	"	78	1,750	24.00	1.50	1,000	1,800
6-pdr Wiard	2.56	steel	56	600	6.00	0.60	1,300	1,800
10-pdr Wiard	3.00	"	58	790	10.00	1.00	1,230	1,850
12-pdr Whitworth ..	2.75	steel & iron	84	1,000	12.00	2.00	1,600	3,000
12-pdr Blakely	3.40	"	59	800	10.00	1.00	1,250	1,850
3-inch Armstrong ...	3.00	"	76	996	12.00	1.25	1,350	2,200
6-pdr Whitworth breechloader	2.15	"	70	700	6.00	1.00	1,550	2,750
12-pdr Whitworth breechloader	2.75	"	104	1,092	12.00	1.75	1,500	2,800
3-inch Armstrong breechloader	3.00	"	83	918	12.00	1.25	1,300	2,100

Characteristics of Principal Civil War Smoothbore and Rifled Field-Artillery Weapons (reprinted from Harold L. Peterson, *Notes on Ordnance of The American Civil War* by courtesy of the American Ordnance Association)

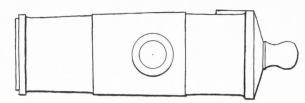

8-INCH SIEGE HOWITZER, MODEL 1841

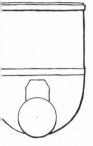

8-INCH SIEGE MORTAR, MODEL 1841

24-POUNDER GUN ON SIEGE CARRIAGE

8-INCH SIEGE MORTAR, MODEL 1861

CONFEDERATE CREW WITH 10-INCH SIEGE MORTAR. FROM A PHOTOGRAPH

COEHORN MORTAR

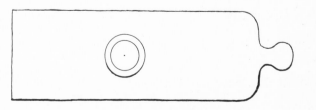

8-INCH SIEGE HOWITZER, MODEL 1861

Siege and Garrison Artillery

SIEGE AND garrison artillery comprised a group of guns, howitzers, and mortars designed for battering an enemy's fortifications, besides defending fixed field positions and in some instances the land faces of permanent forts. In actual use, however, they had no monopoly on these functions. Field pieces sometimes found their way into fixed fortifications, and huge seacoast pieces frequently took position to pound the enemy's defences. In the operations against Fort Sumter, for instance, the Union forces employed the "Swamp Angel," a 200-pounder Parrott which was far bigger than any piece in the official siege and garrison classification, and at the siege of Petersburg they mounted a 13-inch seacoast mortar known as "the Dictator" on a railroad car to bombard the Confederate lines. Even naval guns sometimes served in field forts.

The artillery system of 1841 listed eight principal weapons in the siege and garrison category. These included 12-, 18-, and 24-pounder guns, 8-inch and 24-pounder howitzers, 8- and 10-inch mortars, and a special little 24-pounder coehorn mortar. All except the bronze coehorn were iron pieces. In 1861 new, simpler, more streamlined models replaced the 8- and 10-inch mortars and the 8-inch howitzer, and two new rifles appeared on the lists—the 4½-inch Ordnance rifle and the 30-pounder Parrott.

Siege Gun Carriages

For siege work the guns, the rifles, and the 8-inch howitzer mounted stock trail carriages generally similar to those used for field pieces. They were a good deal heavier, however, and they also had one other distinction. Because of the weight of the tubes they could not be moved safely in their firing position. Like the big Gribeauval pieces they needed a special position on the cheeks for travelling. Instead of a separate set of trunnion notches, however, they boasted curved travelling trunnion beds against which the trunnions could be rested. When in this position the breech was supported by a heavy curved bolster bolted to the top of the trail. The 8-inch howitzer had a special variation all its own. Because its tube was so short, it could not use an elevating screw and so had to be elevated with wedges in the old fashion.

For garrison duty the guns might be mounted on stock trail carriages or sometimes on barbette types to fire over walls. The 24-pounder howitzers were primarily flank defence pieces for big masonry forts, and so they were normally mounted on a modified form of casemate carriage. Mortars, of course, mounted on beds, heavy-cheeked types for the 8- and 10-inch models and solid blocks with carrying handles for the coehorns which could be picked up and carried by four men.

Big Guns in the Civil War

Siege and garrison artillery played an important and sometimes spectacular role in the Civil War. It was a siege mortar that fired the opening shot against Fort Sumter and commenced hostilities. A 30-pounder Parrott rifle accompanied McDowell's army into the field and fired the first shot at Manassas, the first land battle of the War. At Malvern Hill the 1st Connecticut heavy artillery fired 4½-inch Ordnance rifles along with 30-pounder Parrotts in another field battle. At Fredericksburg 4½-inch rifles pounded Marye's Heights, and at Gettysburg the Confederates had at least one big 24-pounder gun in operation. These big pieces, dragged by ten horses were the largest guns that could be hauled over normal roads. Yet here it was on the Confederacy's deepest penetration of the North. Siege and garrison pieces protected Washington and Richmond throughout the War, and at Petersburg they

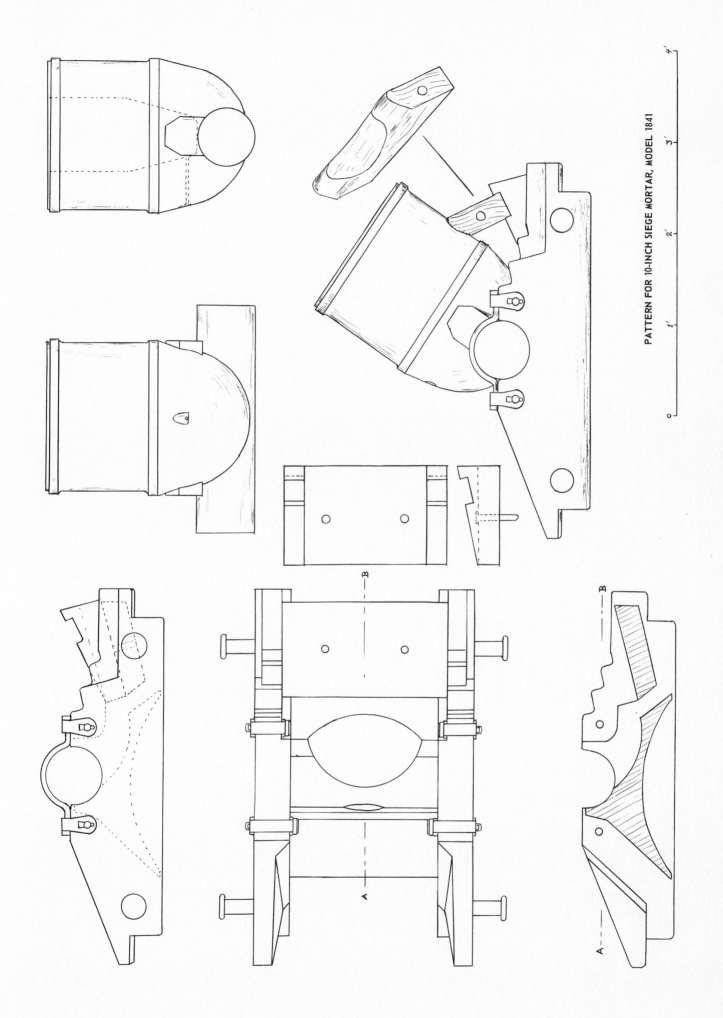

PATTERN FOR 10-INCH SIEGE MORTAR, MODEL 1841

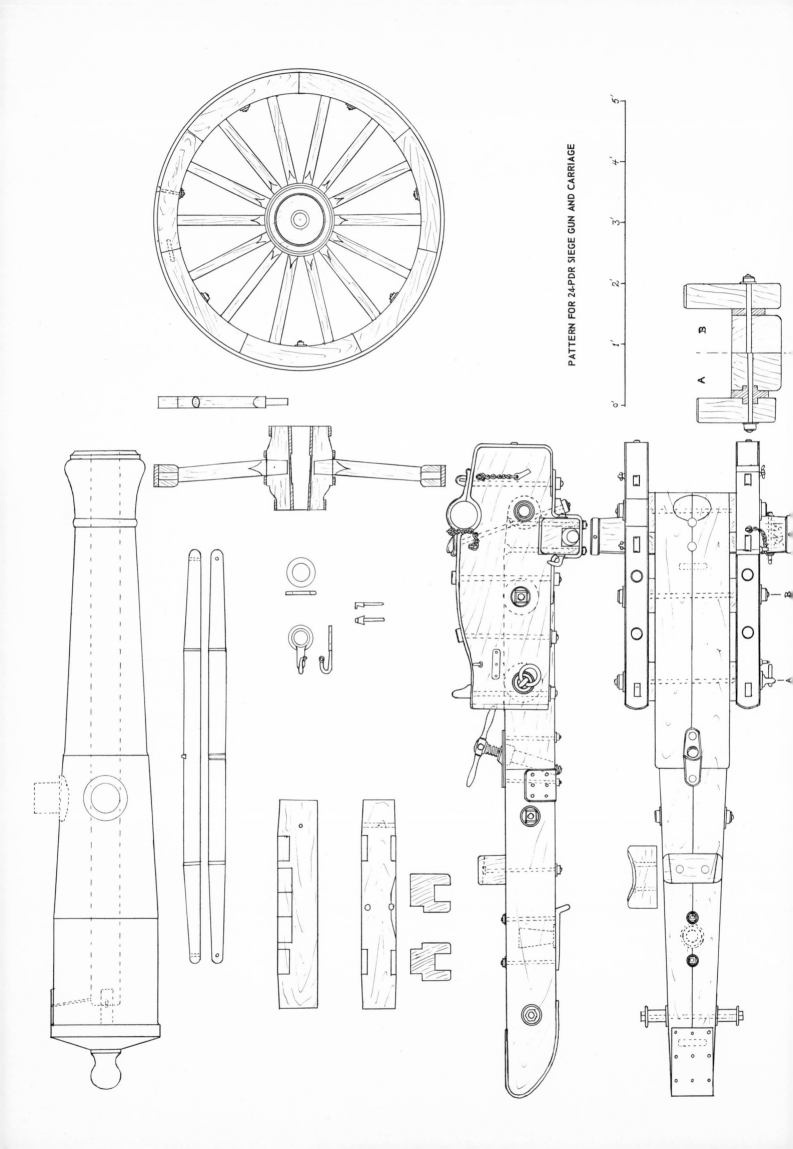

PATTERN FOR 24-PDR SIEGE GUN AND CARRIAGE

boomed their destruction from both sides through the conflict's longest siege. In the West they fought at such field battles as Shiloh as well as in the defenses of Fort Donelson and Vicksburg and other river fortifications. Some even took to the water and fought on the Union river gunboats. Their role was never decisive, but the "heavies" were there from the very beginning to the bitter end.

Developments in Seacoast Artillery

THE CATEGORY of seacoast artillery saw far more development and change than siege and garrison artillery, and it made a far more important impact upon warfare in general than the previous classification. The official registry of pieces designated by Alfred Mordecai in 1841 listed only two guns—32- and 42-pounders—supplemented by 8- and 10-inch seacoast howitzers and 10- and 13-inch seacoast mortars. These were all considered the model of 1841. Of the whole lot, only the two guns remained throughout the period, and even some of them underwent substantial alterations.

The New Columbiads

The first major change to the system of seacoast artillery occurred in 1844, just three years after the system had been adopted. At that date Col. George Bomford, Chief of Ordnance, approved two new columbiads. There had perhaps been columbiads after 1820 and before 1844. The records are replete with mentions of "old model" columbiads and the like, but the picture is not clear. In any event in 1844 two specific columbiads entered the seacoast defences of the United States, and they were officially recognized as such. Like the earlier pieces they seem to have been a combination of both gun and howitzer. They were chambered at the base of the bore like a howitzer, and they threw shells like a howitzer. Yet they could also hurl solid shot like a gun. They seem to have supplanted the seacoast howitzers immediately, for only a very few of these interesting pieces were ever made. But they never superseded the 32- and 42-pounder guns. In 1858 these columbiads were redesigned again on scientific principles that called for the removal of the useless base ring and the equally obsolete muzzle swell. Neither of these traditional features had any effect on a cannon's performance, and they were well omitted. The chamber also disappeared for less obvious reasons, but again only a few of the new model columbiads got into production because a still more efficient piece, the Rodman of 1861, came along just three years later.

Before leaving the columbiad it might be well to mention one further distinguishing feature of this short-lived class of ordnance. It had a ratchet device for elevation that ran up the face of its breech—right through the center of its cascabel knob. This was an important design element, for it permitted an elevation of 39 degrees as compared with 15 degrees possible for a normal gun with an elevating screw. A piece of ordnance designed to throw shells needed this extra area of elevation for high-angle fire, and this made the new columbiads far superior to the older seacoast howitzers.

The Revolutionary Rodmans

The model 1858 columbiads were officially superseded by a new model in 1861. These new pieces first continued the designation of columbiad but soon came to be called Rodmans after their inventor, the distinguished ordnance officer, Thomas Jefferson Rodman. Here was a radically new cannon. It differed from all its predecessors both in external form and in manufacturing technique. Taking the latter aspect first, Rodman pioneered a new production method by which newly cast cannon were cooled from the inside out instead of vice versa. This meant that each succeeding layer was shrunk tightly upon its inner neighbor as it cooled, thus making a cast iron gun much stronger than it had been under the old system of cooling from the outside in.

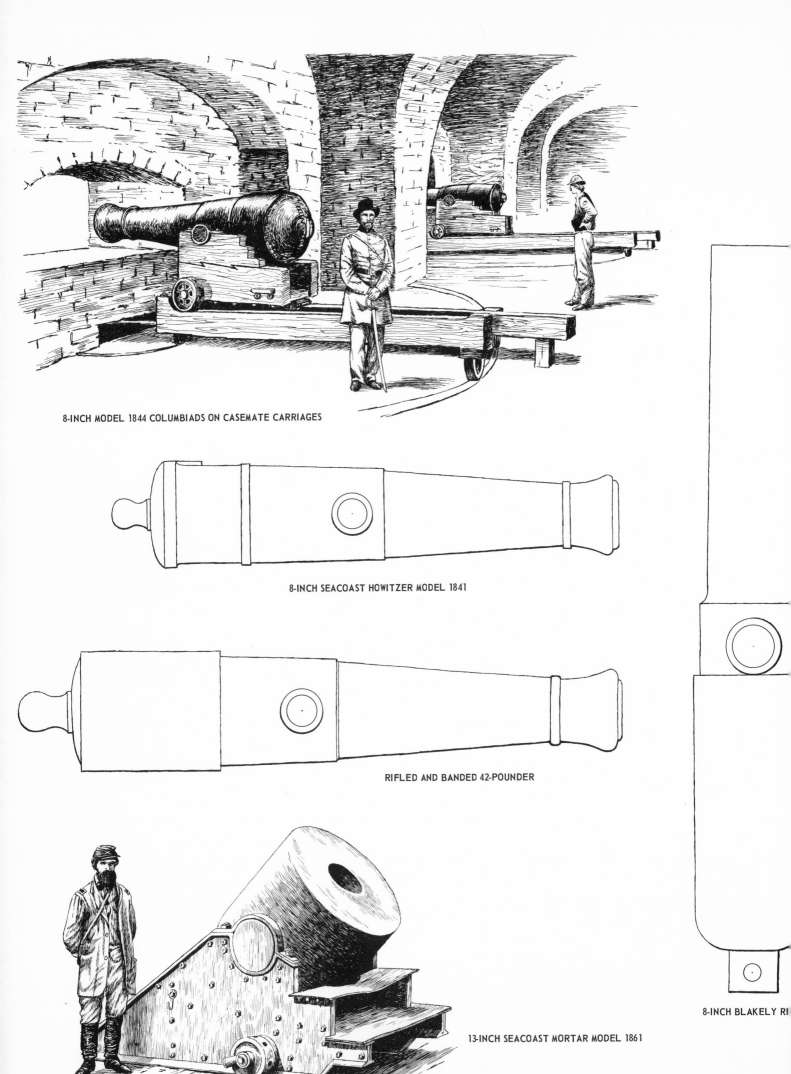

8-INCH MODEL 1844 COLUMBIADS ON CASEMATE CARRIAGES

8-INCH SEACOAST HOWITZER MODEL 1841

RIFLED AND BANDED 42-POUNDER

13-INCH SEACOAST MORTAR MODEL 1861

8-INCH BLAKELY RI

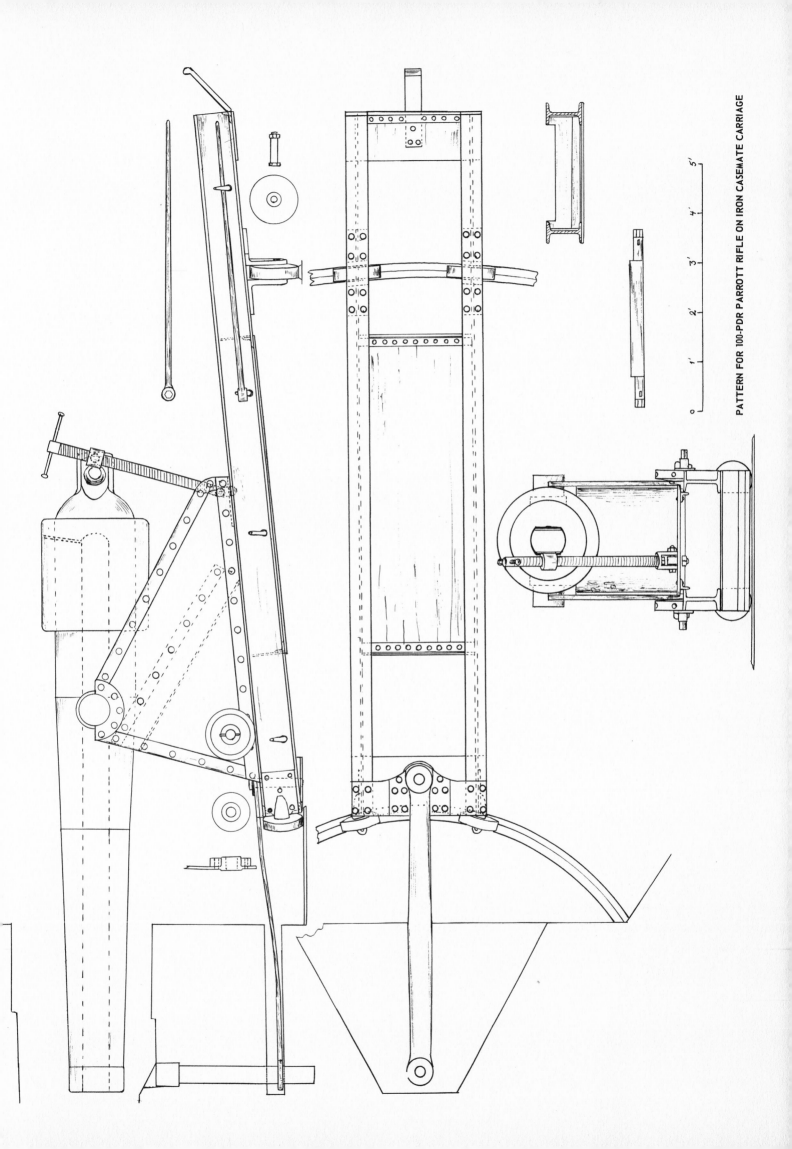

PATTERN FOR 100-PDR PARROTT RIFLE ON IRON CASEMATE CARRIAGE

0   1'   2'   3'   4'   5'

With this new strength, Rodman also designed his pieces in accordance with a curve of pressure determined by the power exerted by powder charges in the bore. He did away with all mouldings, and followed this curve literally so that his guns developed a very streamlined, almost bottle shape. At the same time, however, he retained the ratchet elevating mechanism because of the greater range of positions it gave his pieces. This was a wonderful new weapon. Union artillerymen were quick to grasp its significance. They ordered guns of the new design in huge quantity, and so did the Confederates, although some manufacturers such as Tredegar refused to accept the internal cooling system. Rodmans appeared in short order both North and South. There were 8-inch Rodmans, 10-inch Rodmans, 15-inch Rodmans, and eventually two whopping 20-inch Rodmans, the biggest guns ever cast in America even if they never did see active service. The standard Rodmans were so good that they continued in use, sometimes with experimental breeches, up until the beginning of the present century.

New, Streamlined Mortars

The mortars also underwent change, even if it was less than the columbiads. The old models of 1841 were superseded in 1861 by new 10- and 13-inch versions that eliminated the muzzle rings and other moulding and produced heavier, more streamlined products. These new model 1861's appeared too late to find their way into many of the seacoast fortifications, but they did ride on the mortar ships that plied the Tennessee and Mississippi rivers and dropped their bombs into Confederate strongholds. Most spectacular of all was one special 13-inch model called "the Dictator" that took up a position on a railroad flatcar and participated in the siege of Petersburg. Local tradition has it that the first shell from this monster fell in the dooryard of a house of ill repute between the lines and caused hurried exits from various Confederate officers then frequenting the establishment and who cared little whether it was a door or window that gave them egress. Possibly other model 1861 seacoast mortars performed more effectively, but if so their exploits remain unrecorded.

Giant Rifles

The two guns remained unchanged until 1861, when they were joined by a series of big rifles.

Parrott rifles like the smaller versions fighting in the field joined the seacoast ranks, but these were huge monsters—200- and 300-pounders, in fact, with bores of 8 and 10 inches respectively. Whitworth, Armstrong, and Blakely rifles in bores up to 12¾ inches also appeared on the Confederate side to reply to these Northern giants. The big Parrotts had an unfortunate tendency to blow off their muzzles because of the premature explosion of their shells, and so some artillerymen were less than enthusiastic about them. Yet the great range, accuracy, and penetrating power of the rifle, especially against masonry fortifications, commended them to all artillery and engineer officers. Thus attempts were made to rifle the old smoothbores. In the North the 32's and 42's were rifled and wrapped around the breech with wrought iron bands when necessary for extra strength. The Confederates seem to have rifled and banded almost anything that was serviceable, for one finds pictures and surviving specimens of a whole host of banded and rifled guns even including 8-inch Rodman types.

Carriages and Beds for Seacoast Guns

These big seacoast pieces were mounted on a variety of carriages and beds. In some instances the guns, columbiads, and rifles might be mounted on casemate carriages and fired from the interior rooms of forts. More often they appeared on barbette carriages, firing over the embankments of dirt or masonry emplacements. Sometimes they had front pintle carriages with the pivot at the base of the wall. In other instances they had center pintle mounts with the pivot directly under the gun and four wheels on a track that permitted a wide angle of traverse. Wooden carriages were standard at first, and indeed wooden carriages continued in use throughout the Civil War, especially in the South. But wrought iron carriages received official approval in 1861, and they appeared for both casemate and barbette emplacements in the North whenever possible. They were, after all, a good deal stronger and less bulky than the wooden models. Mortars also switched to light wrought iron beds with steps in front in 1861.

Military Capabilities of Seacoast Guns

These were big pieces, and they were extremely powerful. Yet they were slow. The huge projectiles took at least two men to lift and load on the

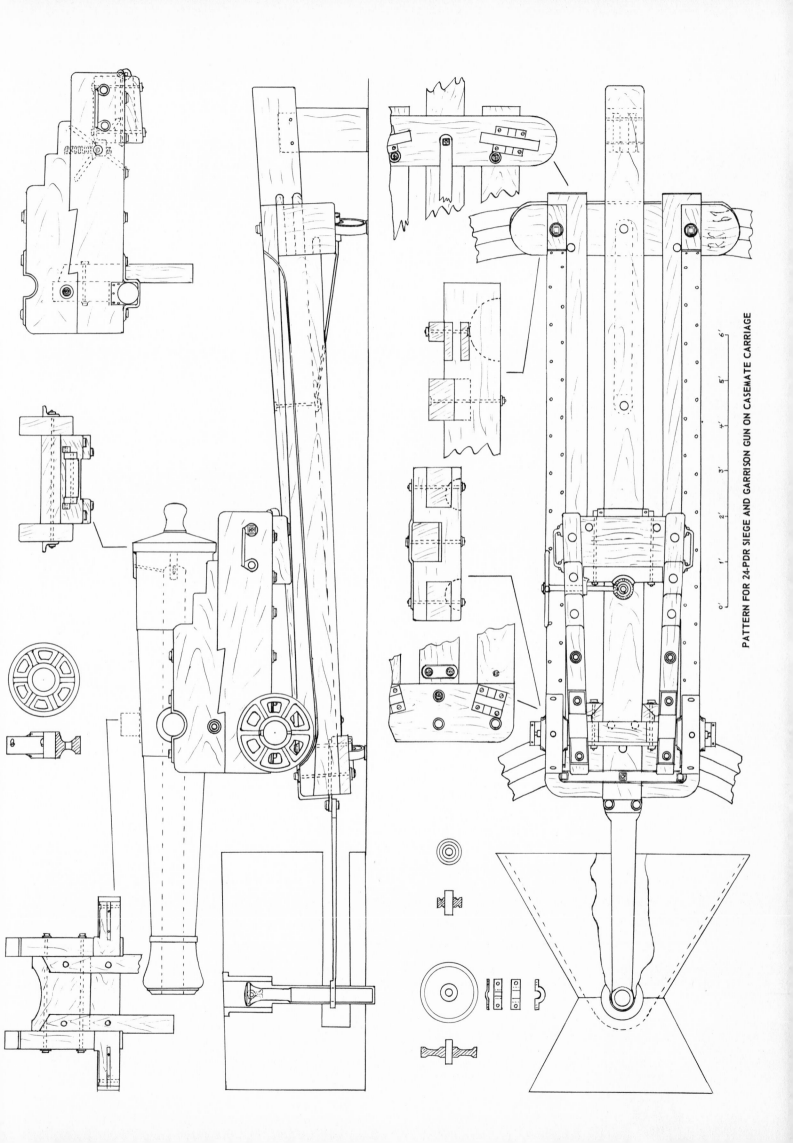

PATTERN FOR 24-PDR SIEGE AND GARRISON GUN ON CASEMATE CARRIAGE

0' 1' 2' 3' 4' 5' 6'

SIEGE AND GARRISON ARTILLERY

	Bore diameter (inches)	Material	Length of tube (inches)	Weight of tube (pounds)	Weight of projectile (pounds)	Weight of charge (pounds)	Range at 5° elevation (yards)
4½-inch rifle	4.50	iron	133.00	3,450	33.0	3.50	2,078
30-pdr Parrott rifle	4.50	"	136.00	4,200	29.0[1]	3.75	2,200
24-pdr gun	5.82	"	124.00	6,240	24.4	6.00	1,901
18-pdr gun	5.30	"	123.25	4,680	18.5	4.50	1,592
12-pdr gun	4.62	"	116.00	3,120	12.3	4.00	1,834
8-inch howitzer	8.00	"	61.50	2,614	50.5[1]	4.00	1,241
8-inch mortar	8.00	"	22.50	930	44.5[1]	3.75	1,200[2]
10-inch mortar	10.00	"	28.00	1,852	87.5[1]	4.00	2,100[2]
24-pdr coehorn mortar ...	5.82	bronze	16.32	164	17.0[1]	0.50	1,200[2,4]

SEACOAST ARTILLERY

	Bore diameter (inches)	Material	Length of tube (inches)	Weight of tube (pounds)	Weight of projectile (pounds)	Weight of charge (pounds)	Range at 5° elevation (yards)
32-pdr gun	6.40	iron	125.20	7,200	32.6	8.00	1,922
42-pdr gun	7.00	"	129.00	8,465	42.7	10.50	1,955
8-inch columbiad	8.00	"	124.00	9,210	65.0	10.00	1,813[5]
10-inch columbiad	10.00	"	126.00	15,400	128.0	18.00	1,814[6]
15-inch columbiad	15.00	"	182.00	50,000	302.0[1]	40.00	1,518[7]
100-pdr Parrott	6.40	"	151.00	9,700	100.0	10.00	2,247[8]
200-pdr Parrott	8.00	"	159.00	16,300	175.0	16.00	2,000
300-pdr Parrott	10.00	"	173.00	26,500	250.0	25.00	—
10-inch mortar	10.00	"	46.00	5,775	87.5[1]	10.00	4,250[2]
13-inch mortar	13.00	"	53.00	17,120	220.0[1]	20.00	4,325[2]
80-pdr Whitworth rifle ...	5.00	iron & steel	120.00	8,960	70.0	12.00	7,722
70-pdr Armstrong rifle breechloader	6.40	"	110.00	6,903	71.7	10.00	2,266[9]
8-inch Blakely rifle	8.00	"	156.00[3]	17,000	200.0	50.00	—
150-pdr Armstrong rifle ...	8.50	"	129.75	14,896	c.150.0	20.00	—
12¾-inch Blakely rifle ...	12.75	"	192.00	54,000	700.0[1]	—	—

[1] Weight of shell.
[2] Mortar ranges are given at an elevation of 45°.
[3] Bore length only.
[4] Designed to be moved and operated by two men.
[5] Obtained ranges of over 4,812 yards with shell and 27° elevation.
[6] Obtained ranges of over 5,600 yards with shell and 39° elevation.
[7] Obtained ranges of 4,680 yards with a 315-lb. shell and 50 lbs. powder at 25°.
[8] Extreme range 8,428 yards.
[9] Muzzle-loading Armstrongs had practically identical dimensions and ranges.

Characteristics of Principal Civil War Siege, Garrison, and Seacoast Artillery Weapons (reprinted from Harold L. Peterson, *Notes on Ordnance of The American Civil War* by courtesy of the American Ordnance Association)

smallest models, and bigger ones required a mechanical shell hoist. Two men frequently had to handle the rammer. It took seven men one minute and ten seconds to load a 15-inch columbiad and two minutes and twenty seconds to traverse it through an arc of ninety degrees.

Despite their sluggishness the great guns were important to the progress of the war. Big rifled smoothbores knocked down the walls of Fort Pulaski in a few hours and utterly ruined a fort that had been considered impregnable. Other seacoast rifles reduced the walls of Sumter to rubble. They brought the day of masonry fortifications to an end. They made the life of a blockade runner miserable, and, as noted above, "the Dictator" produced startled reactions in the siege of Petersburg. It was a notable record that these big guns piled up, totally aside from the fact that they gave the United States one of the most strongly defended coastlines in the world.

The Great Leap Forward in Ammunition

AMMUNITION SAW some of the most startling developments of any phase of artillery during the years from 1836 to 1865. The old smoothbore types continued relatively unchanged, but it was really projectile innovations that permitted the effective rifles of the Civil War, and accompanying them came new fuzes that offered more versatility in their employment. For years there had been only gradual change at the most and more often complete stagnation in artillery ammunition. Now came the great leap forward.

Replacement of Grape by Cannister

In the more conservative smoothbore projectiles the old forms originally designed by Gribeauval remained supreme. Fixed ammunition with solid shot, simple shell, and cannister along with the new spherical case ruled the field. The major change outside a new fuze was the elimination of grape shot from the repertoire of field guns. At least from 1841 on, the only antipersonnel loads for field pieces were spherical case and cannister. The change made good sense. Spherical case reached the enemy at greater ranges than either grape or cannister could hope for, and at close range cannister offered far more shot than grape. It was also easier on the bore of the guns. Every artilleryman understood the change and the reasons for it, but these facts were completely lost on infantry and cavalry. To the end of the muzzle-loading period every non-artillerist who wrote his memoirs insisted that he had been shot at with grape and spoke fondly of the "hails of grape" that had mercifully spared his life as he charged enemy cannon. Some of their mistaken statements actually became classics of American war prose. Take the famous quote of General Zachary Taylor at Buena Vista when he supposedly gave the order:

"A little more grape, Captain Bragg!" now quoted by every historically minded schoolboy. Alfred Pleasonton, who was there and overheard the actual conversation, reported more logically that it really went like this:

"What are you firing, Captain?"
"Cannister, sir."
"Double or single?"
"Single, sir."
"Then double it and give 'em Hell!"

This, any artilleryman would believe. But tradition dies hard.

Smoothbore Fuzes

The new fuze that brought change to smoothbore ammunition was the invention of a Belgian, a Captain Bormann. It was generally a disc-shaped plug of white metal that could be screwed into the shell. Its major feature was a train of powder in the form of an arc that led to the powder charge within the shell. This arc was marked off on the exterior into quarter-second intervals with a total burning time of five seconds. The cannoneer setting the fuze simply punctured the train at the proper spot for the length of time he wanted the

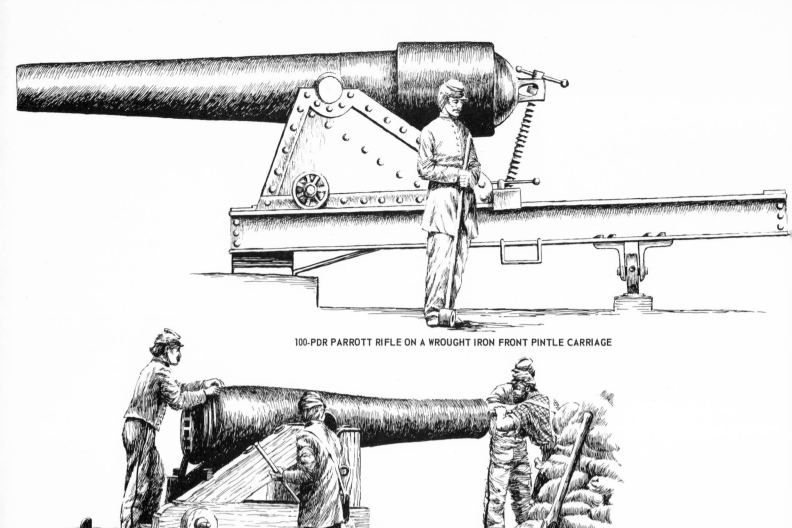

100-PDR PARROTT RIFLE ON A WROUGHT IRON FRONT PINTLE CARRIAGE

8-INCH RODMAN ON A WOODEN CENTER PINTLE BARBETTE CARRIAGE

NAVAL GUN ON A FRONT PINTLE WOODEN BARBETTE CARRIAGE

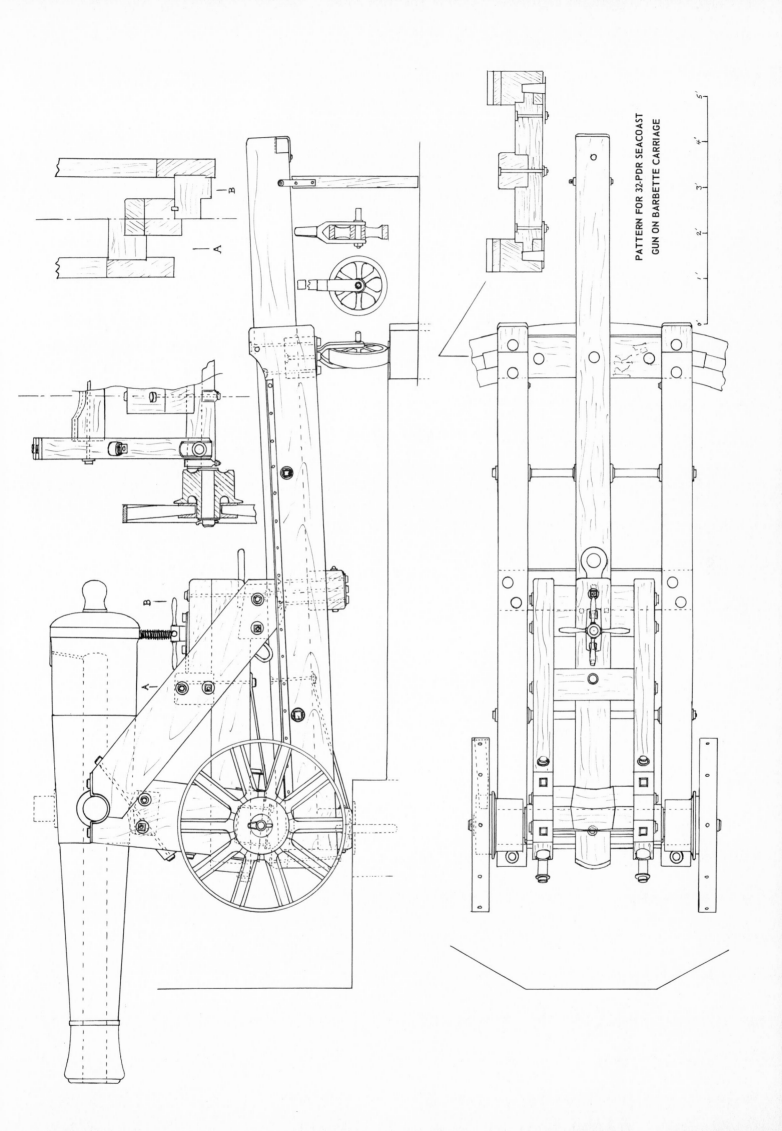

PATTERN FOR 32-PDR SEACOAST
GUN ON BARBETTE CARRIAGE

fuze to burn before exploding the shell, and the flash from the propellant charge in the cartridge ignited it at that point as he wished. It was a simple and very effective fuze. Both the Union and Confederate armies used it for field guns and howitzers during the Civil War, though the Confederates frequently had trouble in obtaining a sufficient quantity of Bormann fuzes and so reverted on occasion to the simpler paper fuze described below.

Some shells, however, were not suited to the Bormann fuze. These were the big mortar shells and columbiad shells that traveled great distances and so needed fuzes that burned longer than five seconds. These big shells continued to use the older wooden fuzes which consisted of a tapered wooden plug containing a hole filled with quick match or portfire composition in its center. The outside of these plugs was marked off with lines indicating the burning speed of the composition within, and so the cannoneers set them by sawing them off at the proper length for the time duration they needed.

Fixed and Semi-fixed Rounds

All smoothbore field piece ammunition was fixed with the projectile strapped to a sabot, and the sabot in turn tied fast to a flannel cartridge bag containing the powder charge. Bigger guns usually used semi-fixed ammunition with the projectile and powder bag separate. Fixed ammunition for field guns remained relatively the same diameter for both shot and powder bag. For howitzers the sabot tapered, and the powder bag was narrower so that it would fit into the chamber of the piece.

Rifle Projectiles

The smoothbore types were the only forms in service through the Mexican War. Then, just before 1860, the first rifled projectiles appeared on the scene. The primary problem to be surmounted in designing rifled projectiles was to obtain some contact with the twisting rifling grooves of the barrel so that the shot would spin in flight and obtain the benefits of the rifling. At the same time the shot could not fit so tightly that they would be difficult to load. Various designers tackled the problem from a number of directions. Some, like one of the Blakely types and one of the Armstrongs, boasted studs on the sides of the projectile

that had to index with the rifling. The Whitworth with its hexagonal flats that corresponded with the flats of the rifling employed a similar principle. All of these suffered from friction in loading and from the need to fit them accurately into the bore so that the studs or flats matched properly. They also tended to require precision manufacture and hand work that made them expensive. The Confederates even tried one projectile with wings so that it would spin in flight even when fired by a smoothbore, but it too was expensive to manufacture and never popular.

The most successful types of rifle projectiles were those which expanded upon firing so that they could be loaded loosely yet fit tightly as they left the bore. One of the first such projectiles in the United States was the Reed, which had a wrought iron cup cast into the base. This cup was pushed outwards by the gases from the propellant charge and took the rifling much in the way that the hollow-based minié ball operated for small arms. It was eminently practical and supplied the ammunition for the early Parrott rifles until it was supplanted by a newer Parrott version with a brass or copper rotating band cast onto the base that expanded in a similar manner. Many Confederate projectiles continued to employ Reed-type projectiles with brass or copper cups on the base, and they are usually classified today as "Confederate Parrott" projectiles although Confederate Reeds would be more accurate. In the North the Reed and Parrott projectiles were used almost entirely in Parrott rifles. The South seems to have employed them somewhat more widely.

Other projectiles with expanding cups or bands of soft metal such as brass, copper, or lead included the Confederate Burton, the Union Dyer, and the Confederate Mullane. This last was distinguished by a rotating disc of brass or copper that was bolted fast and held in position by three dowels instead of being cast in place. One form of the English Armstrong projectile also used a lead coating over a ribbed iron core.

There were other expanding systems as well. One of the most popular in the North was the Hotchkiss, which was fired in the Ordnance rifles. These well-designed projectiles consisted of two iron elements with a lead band between them. Firing forced the two iron pieces together and spread out the lead to take the rifling. A somewhat similar idea was employed in the Confederate Archer, but in this case the lower element was a wooden sabot, and both the sabot and the lead

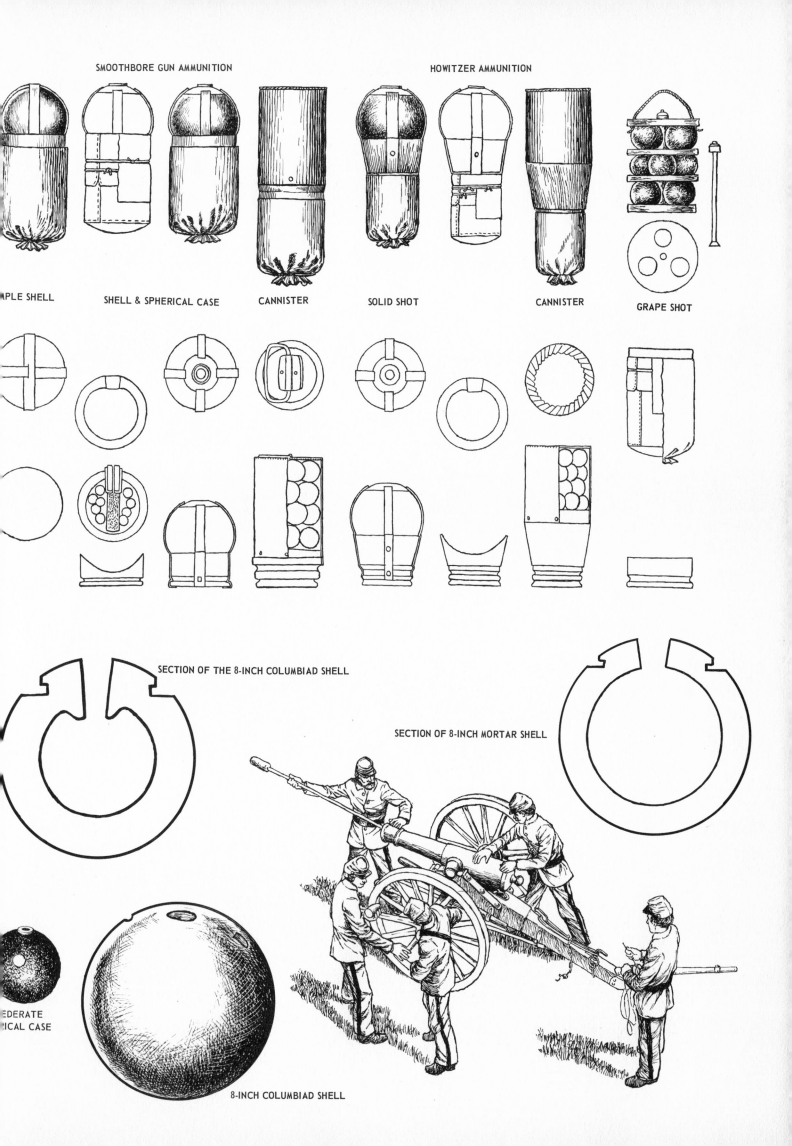

SMOOTHBORE GUN AMMUNITION

HOWITZER AMMUNITION

PLE SHELL

SHELL & SPHERICAL CASE

CANNISTER

SOLID SHOT

CANNISTER

GRAPE SHOT

SECTION OF THE 8-INCH COLUMBIAD SHELL

SECTION OF 8-INCH MORTAR SHELL

EDERATE
ICAL CASE

8-INCH COLUMBIAD SHELL

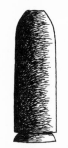

PARROTT-REED
SHELL WITH BASE
EXPANSION CUP

PARROTT SHELL
WITH EXPANSION RING

HOTCHKISS SHELL, THE IRON ELEMENTS (LEFT), THE
COMPLETE PROJECTILE WITH LEAD EXPANSION RING (RIGHT)

SCHENKL SHELL, THE IRON PART (LEFT),
THE COMPLETE PROJECTILE WITH PAPIER
MACHE SABOT (RIGHT)

CONFEDERATE
REED SHELL

CONFEDERATE ARCHER BOLT, THE IRON ELEMENT (LEFT), THE LEAD ROTATING
BAND IN PLACE AND THE WOODEN SABOT BELOW (CENTER), AND THE COMPLETE
PROJECTILE WITH TINNED   IRON CASING OVER THE ROTATING BAND AND SABOT

CONFEDERATE BURTON BOLT, THE IRON PART
(LEFT), WITH LEAD SABOT (RIGHT)

CONFEDERATE
MULLANE SHELL

JAMES BOLT, THE IRON PART (RIGHT), THE COMPLETE
PROJECTILE WITH LEAD AND CANVAS WRAPPING (LEFT)

CANNISTER FOR RIFLE

SAWYER SHELL

ARMSTRONG SHE
WITH LEAD COATI

ARMSTRONG WITH
LEAD COATING REMOVED

ARMSTRONG SHELL
WITH STUDS FOR
RIFLING

WHITWORTH BOLT

DYER SHELL WITH LEAD
EXPANSION RING

8-INCH BLAKELY BOLT

CONFEDERATE WIN
PROJECTILE

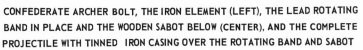

band were enclosed in a tinned iron casing. The Schenkl projectile used a papier-mâché sabot that slid up an expanding shank with the force of the charge and so took the rifling. The Sawyer combined several principles. It had a flanged projectile covered with a lead coating and a copper casing so that it supposedly would expand to fit the grooves. A final approach to the rifling problem was exemplified in the James. This projectile had a hollow base with slanted slots around the sides. According to theory the gases from the propellant charge would enter the base and pass out through the slots. In so doing they would force a lead and canvas wrapping on the outside of the shell into the grooves and so take the proper spin. And there were others.

Fuzes for Rifle Shells

With these spinning projectiles it was possible to know which end would strike the target, and so it became practical for the first time to design a fuze that could be exploded by impact. This had never been feasible with round shot that might strike in any position. Interestingly it was the same men who had created projectiles that gave this new possibility their attention. Charles James was one of the first. He produced a free-floating plunger with a percussion revolver cap on its nose that would be thrown forward by the force of impact and detonate against a screw cap in the nose of the shell. It worked fine, but the trouble was that it took very little impact to set one off, and accidents happened. Schenkl and Hotchkiss followed with similar designs, but Schenkl added a set screw and Hotchkiss a wire that had to be sheared off by the impact before the plunger could hurtle forward and detonate the cap. Thus they offered safety from loading accidents and their fuzes were much more popular with artillerymen. Robert P. Parrott designed quite a different fuze with two thin leaves that had to be twisted off by the spiraling motion of the shell in the bore to free the plunger. Again there were other types

of ingenious percussion fuzes, but these were the most widely used for land artillery during the Civil War.

It was all well and good to have shells explode when they struck their target, but sometimes an artillerist wants his projectile to explode in the air over the head of the enemy. This is particularly true when he is firing case shot. For such uses he needs a timed fuze, and these, too, were available for the new rifled projectiles. For such purposes the cannoneer used a device called a fuze plug, a brass insert that screwed into the fuze hole in the nose of the shell. It presented a smooth cylindrical hole of the proper diameter for one of the various paper fuzes that awaited his selection. These fuzes were simply lengths of quick-burning composition wrapped in paper. The compositions varied in burning time so that there were fuzes available for different lengths of flight, and the burning time was marked both on the fuze itself and the packet in which it came ready to be selected and placed in the fuze plug for firing. In an action it was a matter of seconds to select a prepared shell or case shot and insert the proper paper fuze. As mentioned above, the Confederates used these paper fuzes in their spherical shells as well as in rifled projectiles, but in the North they were employed in rifle projectiles only. Sometimes an umbrella-like metal flash concentrator was used in an attempt to make sure the flash from the charge would ignite the fuze, but these seem to have been a fad and not really necessary.

With his multiplicity of projectiles and fuzes the Civil War cannoneer was better prepared for any possible situation than any artilleryman who had preceded him. In seconds he could select the proper projectile for counterbattery fire, for antipersonnel work at long range or point blank, for ricochet fire, for battering shot, or for any of the other ways in which artillery might be employed. And he could quickly insert the proper fuze for just the result desired. The various choices gave him a mastery of his profession never before equaled, and in four bloody years he made the most of it.

mplements and Instruments

WHEN ALFRED MORDECAI worked out a new system of artillery for the land service of the United States in 1841, he included all the implements and instruments as well

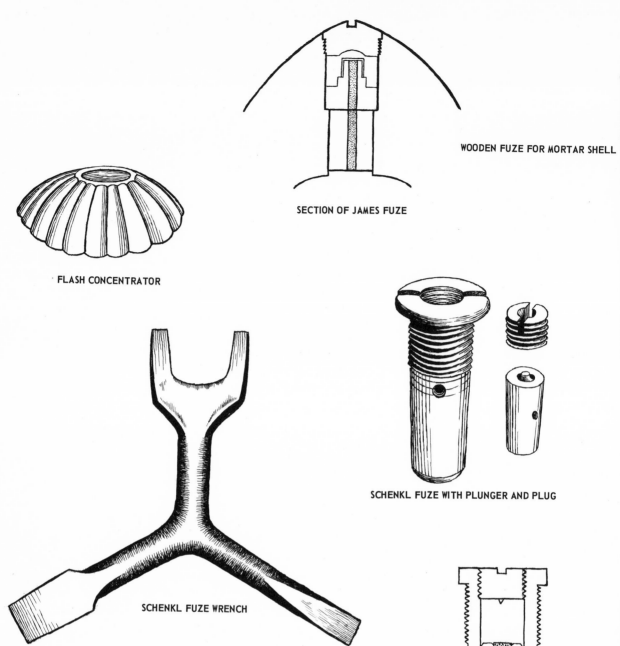

FLASH CONCENTRATOR

SECTION OF JAMES FUZE

WOODEN FUZE FOR MORTAR SHELL

SCHENKL FUZE WITH PLUNGER AND PLUG

SCHENKL FUZE WRENCH

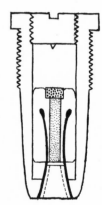

SECTION OF HOTCHKISS FUZE

PARROTT FUZE

FUZE PLUG

PAPER FUZE AND PACKET

BORMANN FUZE

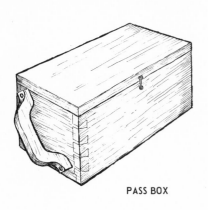

PASS BOX

WATER BUCKET FOR SEACOAST ARTILLERY

LINSTOCK

PORTFIRE HOLDER

BUDGE BARREL

SHELL TONGS

FORK FOR HOT SHOT

as the cannon and carriages. Everything was specified to the smallest detail. And there were innovations. Sponges, rammers, worms, handspikes, pass
boxes, budge barrels, and the like remained the
same in principle. Only minor details were altered.
The big changes came in the method of firing
cannon and in aiming them.

The Friction Primer

Linstocks and portfires continued to be standard items of equipment for every battery, but
now they were emergency gear, designed for use
when the new friction primers ran out or were
otherwise unavailable. These new primers were
a radical improvement over anything seen here-
tofore for firing a cannon. Each primer consisted
of a copper tube with a twisted wire running
through it at right angles and a filling of fuze
composition. A loop at the end of the wire
served for attaching a lanyard, and when this was
pulled it ignited a friction compound somewhat
on the principle of the modern friction match,
set off the fuze compound, and directed a hot
spurt of flame towards the cartridge. Ignition of
the cartridge was almost absolutely certain with
the new primers, and what is more, they with-
stood dampness. In fact, tests showed that friction
primers soaked in water would still usually ignite
and wet ones, when dried out, were absolutely as
good as new. Mordecai probably did not have these
primers ready for his system in 1841, but by the
time his standard work appeared in 1848 they
were an accepted part of the artillerist's regimen.
And they were a welcome part, too. No longer was
it necessary to kindle a length of match or a portfire and keep it burning. The new primers were
ready in an instant, and they were very, very fast
and sure.

New Aiming Devices

The improvements in aiming, too, were important. The gunner in the beginning still used a
level to find the line of sight on a piece that had
its wheels on uneven ground. Then he used a
tangent scale to arrive roughly at the proper elevation for his shot. This was the sighting equip-

ment specified by Mordecai as the most modern,
and it served U. S. artillerymen through the Mexican War. By the time of the Civil War, however,
new sights had appeared that were infinitely more
accurate and easy to use. For smoothbores there
was the pendulum hausse. This device fitted into
brackets mounted at the top of the breech face.
It had miniature trunnions on a pivot and had a
weight at the bottom so that it always assumed an
absolutely vertical position no matter how rough
the ground. The gunner could throw away his
level. A movable bar slid up and down on the body
of the new sight and quickly indicated the necessary degree of elevation. It was a fine, accurate,
quick instrument. Rifles had sights of their own.
The Ordnance rifle used the pendulum hausse like
the smoothbores. Wiard developed a complicated
instrument that mounted on the cascabel knob and
offered intricate adjustments. It probably never
saw service. More common was the Parrott sight,
a brass rod marked off in degrees of elevation
with a movable sighting device that could be
raised and lowered and fastened in position with
a set screw. This sight mounted in a loop on the
right side of the breech where it aligned with a
front sight on the right rimbase.

Lanyards and Tow Hooks

Among the newer items of the more prosaic
implements were shell tongs for carrying heavy
smoothbore projectiles, special forks for hot shot,
and a little iron hammer with a bent and pointed
handle called a tow hook. This implement was
used to pull out the tow packing around shells
when they reached the field in shipping cases, and
hence the name. The hammer end was designed
so that gunners could replace any loose nails on
the straps that held spherical shot and shells to
their sabots. New, too, was the lanyard used by
the number 4 cannoneer to fire the fine new friction primers. This was the first appearance of a
device that is still used by gunners and has become so typical that it is sometimes used as a
distinctive insignia. And many an old artilleryman
speaks fondly of the "Order of the Purple Lanyard."

HANDSPIKE FOR FIELD GUN

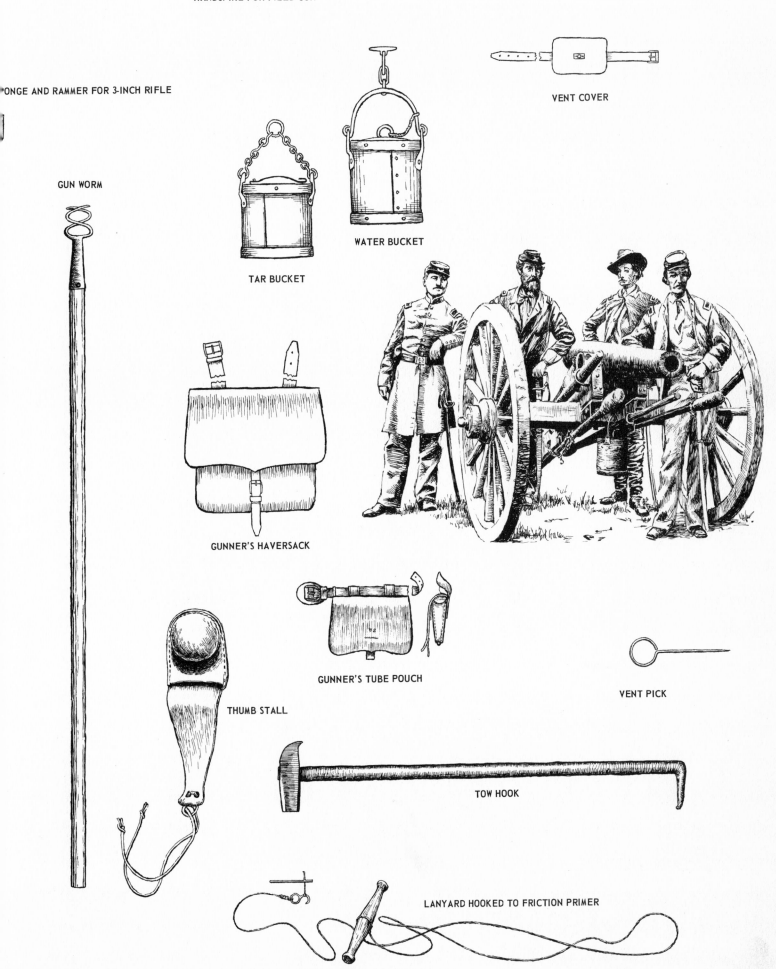

ONGE AND RAMMER FOR 3-INCH RIFLE

VENT COVER

GUN WORM

TAR BUCKET

WATER BUCKET

GUNNER'S HAVERSACK

GUNNER'S TUBE POUCH

VENT PICK

THUMB STALL

TOW HOOK

LANYARD HOOKED TO FRICTION PRIMER

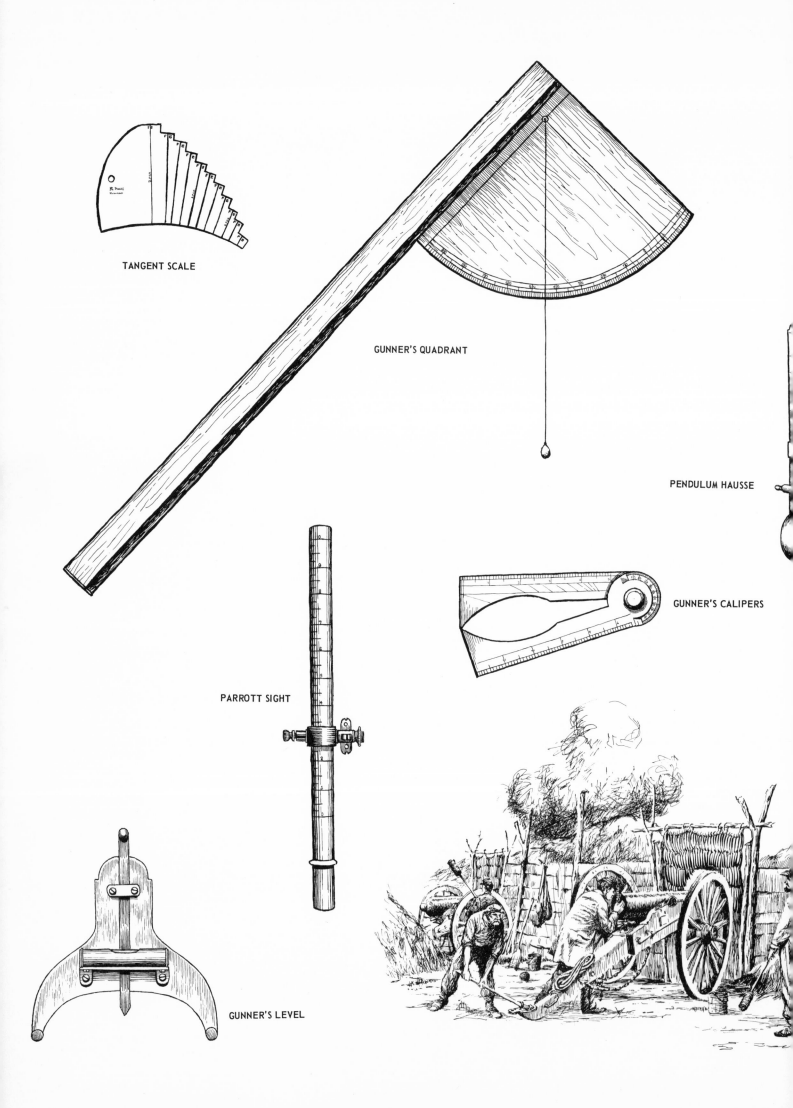

TANGENT SCALE

GUNNER'S QUADRANT

PENDULUM HAUSSE

GUNNER'S CALIPERS

PARROTT SIGHT

GUNNER'S LEVEL

treamlined Gun Drill

BUT THE lanyard was only one of the things that had changed about the gun crew. The old group of unskilled cannoneers or matrosses who had stood by to handle the drag ropes and maneuver the piece on the field disappeared. Whenever the cannon had to be manhandled the basic crew itself now performed that duty. The average field gun was served by a crew of only eight men—six or seven cannoneers and a gunner. The gunner, frequently a noncommissioned officer, pointed the piece and gave the firing command. Number 1 now handled the sponge and rammer. Number 2 loaded. These two worthies maintained their positions on the right and left of the muzzle respectively, but now they stepped outside the wheels at the command "ready" before the piece was fired. Number

3 still thumbed the vent, but he had switched back to the right of the breech and thumbed with his left hand. He also used the priming wire or vent pick to break open the cartridge. Number 4, on the left of the breech, inserted the primer and pulled the lanyard, firing the piece. Number 5 carried the ammunition from number 6 or number 7 at the limber chest to number 2 at the muzzle.

A well-drilled eight-man crew functioned smoothly and efficiently as had its more numerous forebears. Normally a good smoothbore crew firing fixed ammunition was expected to get off two aimed shots a minute with solid balls, shells, or spherical case. At point-blank range with can- nister they could be expected to double this to four shots a minute. But they often did better than this by eliminating some of the standard motions.

The first thing to go was sponging the bore between shots. This was a dangerous motion to eliminate because of the possibility of leaving smoldering remnants of cartridge bag and because it allowed the piece to become very, very hot. Still it was sometimes less dangerous than allowing charging infantry to get too close. In a hot action, number 1 might sponge only every four or five shots, and number 3 would curse as the hot tube scorched his thumbstall and seared his thumb inside it. Augustus "Cub" Buell, who served as number 3, recalled such a situation when Stewart's Battery "charged" the Confederates at Bethesda Church in 1863 and wiped out an enemy battery at almost point-blank range. With typical Victorian reticence he reported his altercation with the num- ber 1 man:

I shall never forget the behavior of our No. 1 in this action. It was old Griff Wallace, of the 7th Wisconsin. He was certainly an artist at the muzzle of a gun. On this occasion he didn't pre- tend to sponge, except at about every fifth load. Meantime the hot vent was burning my thumb- stall to a crisp and scorching my thumb, so I would call out:

"For ——— ———'s sake, Griff, sponge the gun!"

And he would answer:

"Sponge, ——— ———!" "Stick to the vent, you little ——— ———!!"

"Stick!!!"

Ordinarily I would have resented that epithet, but did not feel called upon to do so then. To- ward the last it was really painful. As the leather kept burning through I would pull the thumb- stall down until no more of it was left, and then I appealed to Griff that the vent was burning my flesh. All the satisfaction I got was a fierce growl between his Irish teeth:

"Thumb it with the bone, then, ——— ——— you!!"

I can still see that Irish hero now, his curly hair loose on his bare head, his arms bare to the el- bows, as he had thrown away cap and jacket, and rolled up his shirt sleeves when we unlimbered. After it was all over, and we were sipping our coffee under the shadow of Griffin's headquarters at the little church that evening, I said:

"Griff, suppose I had let go that hot vent when you wouldn't sponge, and there had been a pre- mature discharge in consequence?"

"Well," he says "Cub, I had thought of that, and had made up my mind to brain you at once with the rammer-head if that occurred!"

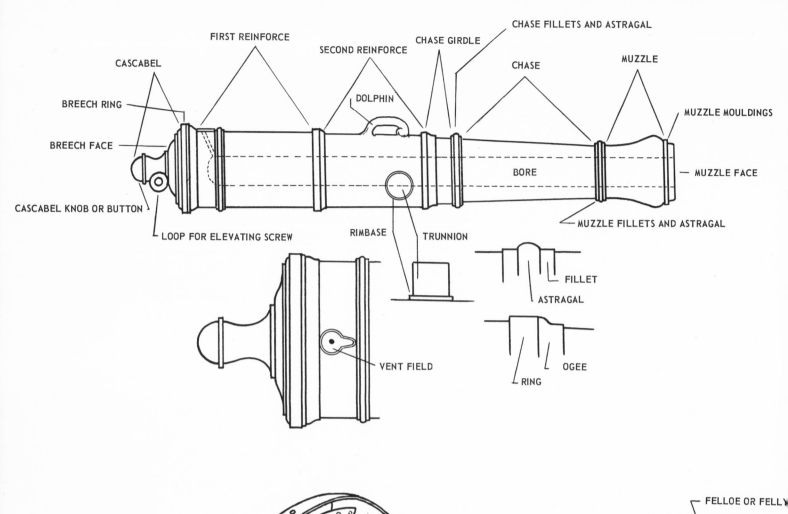

FIRST REINFORCE

SECOND REINFORCE

CHASE FILLETS AND ASTRAGAL

CASCABEL

CHASE GIRDLE

CHASE

MUZZLE

BREECH RING

DOLPHIN

MUZZLE MOULDINGS

BREECH FACE

BORE

MUZZLE FACE

CASCABEL KNOB OR BUTTON

MUZZLE FILLETS AND ASTRAGAL

LOOP FOR ELEVATING SCREW

RIMBASE

TRUNNION

FILLET

ASTRAGAL

VENT FIELD

OGEE

RING

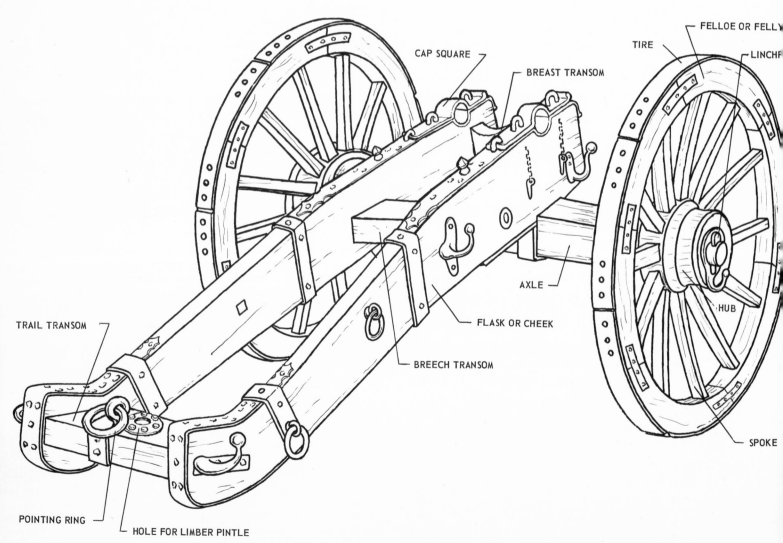

CAP SQUARE

BREAST TRANSOM

FELLOE OR FELLY

TIRE

LINCHP

TRAIL TRANSOM

AXLE

HUB

FLASK OR CHEEK

BREECH TRANSOM

POINTING RING

HOLE FOR LIMBER PINTLE

SPOKE

How deliciously Irish that was! The joke of this will instantly be understood by any artilleryman. If I had ever let go of that vent there wouldn't have been enough left of Pat and his rammer to brain a flea. He would have been blown from the muzzle.

Buell and his friends were firing an old-style Napoleon with dolphins, one of the very first ones made. Since this was counterbattery work they were aiming their shots carefully, yet they undoubtedly got off more than the expected two shots a minute. They had to hold their position. Other types of pieces were not quite so fast. Rifles which fired semi-fixed loads were a trifle slower than the smoothbores because of the extra motions in loading and ramming, but they employed the same number of men in a crew. The crew for a mortar involved three or five men, depending upon its size. They usually figured on getting off about twelve shots an hour, but under emergency conditions they might increase this to twenty shots. Seacoast pieces with their great heavy projectiles required still more time and energy. It frequently required two men to carry even a shell, let alone a solid shot, and often two men were needed to ram it home. Seven men could load and fire a 15-inch columbiad in one minute and ten seconds, but this did not include aiming. This would depend upon the extent of change necessary in elevation and deflection. It took two minutes and twenty seconds to traverse one of these monsters through an arc of 90 degrees. With this slow rate of speed and the swiftness of the new steamships, everything depended upon the accuracy of these big guns. Frequently it was only possible to fire such a piece once at a fast vessel, and so the aim needed to be perfect. When it was the results were impressive as heavy shot weighing from 100 to 300 pounds struck home.

Whether they served fast-firing field pieces or slow-shooting behemoths, the artillerists of the Civil War took their tasks seriously. They braved hails of lead and cannister, strained to carry 300-pound projectiles at the double, seared fingers on hot vents, and sweated to raise their rate of fire to the maximum and keep it there till an action was won. In addition they studied all that was known of the science of gunnery. They memorized aiming tables and fuze settings and learned to estimate distance accurately, all in an effort to make their practice just as good as possible. And they succeeded. The proud tradition that had begun in America with de Soto's anonymous brass pieces of 1539 culminated in 1865 as muzzle-loading artillery reached the very apex of its perfection. Further developments had to await the breechloader and the rocket. Dedicated crews and gifted designers had achieved the utmost with the solid-breeched tubes. An era reached its climax—and came to an end.

Bibliography

PRIMARY SOURCES

The Beginnings

BINNING, CAPTAIN THOMAS, *A Light to the Art of Gunnery*, London, 1677. This was reprinted several times, the last time in 1744.

BLONDEL, FRANÇOIS, *L'Art de Jetter les Bombes*, Paris, 1683.

BOURNE, WILLIAM, *The Arte of Shooting in Great Ordnaunce*, London, 1587.

BOURNE, WILLIAM, *Inventions or Devises*, London, 1578.

BUCHIERS, *Theoria et Praxis Artilleriae*, Nuremberg, 1682.

BUSCA, GABRIELO, *Instruzione dei Bombardieri*, Venice. Several editions between 1545 and 1584.

CAPOBIANCO, ALLESSANDRO, *Corona e Palma Militare di Artigliera*, Venice, 1598.

CERDA, T. DE, *Lecciones de Artilleria*, Madrid, 1644.

COEHORN, MENNO VAN, *Instructions sur le Fait de l'Artillerie*, Paris, 1633.

COLLADO, LUIS, *Platica Manual de Artilleria*, Milan, 1592. This is a greatly revised edition of a work by the same author entitled *Pratica Manuale di Arteglieria*, which appeared in 1586. A number of editions of these two works appeared in both Italy and Spain.

DAVELCOURT, DANIEL, *Trois Traictez sur le Faict de l'Artillerie*, Paris, 1616.

ESSENWEIN, A., *Quellen zur Geschichte der Feuerwaffen*, Leipzig, 1877.

GAYA, LOUIS DE, *Traité des Armes*, Paris, 1678. Reprinted in facsimile, London, 1911.

GENTILINI, EUGENIO, *Instruttione de' Bombardieri*, Venice, 1592.

HASSENSTEIN, WILHELM, editor, *Das Feuerwerksbuch von 1420*, reprint edition, Munich, 1941. Contains pictures of contemporary cannon as well as some from the early 16th century.

HEXHAM, HENRY, *The Third Part of the Principles of the Art Military*, Rotterdam, 1643.

MOORE, SIR JONAS, *A General Treatise of Artillery*, London, 1683. This is primarily a translation of Moretti, *Trattato Dell' Artiglieria*, 1672.

MORETTI, TOMASO, *Trattato Dell' Artiglieria*, Brescia, 1672.

NORTON, ROBERT, *Of the Art of Great Artillery*, London, 1624.

NORTON, ROBERT, *The Gunner Shewing the Whole Practice of Artillery*, London, 1628. This is almost an exact translation of Ufano and reproduces his plates.

RIVAULT, DAVID, Sieur de Fleurance, *Les Elemens de l'Artillerie,* Paris, 1608.

ROBERTS, JOHN, *The Compleat Cannoniere,* London, 1639. Especially good for dimensions, charges, and ranges.

SARDI, PIETRO, *L'Artigliera,* Venice, 1621.

SIEMIENOWICZ, CASIMIR, *Ars Magna Artilleria,* Amsterdam, 1650. This was a widely popular work with editions in English, French, Italian, and German, some appearing as late as 1729.

SMITH, THOMAS, *The Art of Gunnery,* London, 1600.

SMYTHE, SIR JOHN, *The Whole Art of Gunnery,* London, 1628.

TARTAGLIA, NICCOLO, *Quesiti et Inventioni Diverse,* Venice, 1528.

UFANO, DIEGO, *Artillerie,* Zutphen, 1621. This work appeared with different titles in Spanish, Italian, French, Polish, and German from 1613 to 1643. It was a highly influential work. An English version appeared as Robert Norton, *The Gunner Shewing the Whole Practice of Artillery,* 1628, and again in William Eldred, *The Gunner's Glasse,* 1646.

WINKRAT, *Compendium der Artillerie,* Innsbruck, 1685.

The French Wars and the Revolution, 1689-1783

Compendio de Artilleria, Cadiz, 1754.

DIDEROT, DENIS, *Encyclopédie ou Dictionnaire des Sciences, des Arts et des Metiers,* 17 volumes, Paris, 1751-1777. There is an especially good pictorial section on the manufacture of cannon.

GRIBEAUVAL, JEAN BAPTISTE, *Tables des Constructions des Principaux Attirails de l'Artillerie,* Paris, 1792. 3 volumes of text plus a huge atlas of plates.

MULLER, JOHN, *Treatise of Artillery,* London, 1757. This appeared in several editions between 1757 and 1780, including a pirated American edition in 1779.

ROBERTSON, JOHN, *A Treatise of Such Mathematical Instruments as are Usually Put into a Portable Case . . . with an Appendix Containing the Description and Use of the Gunners Callipers,* 3rd edition, London, 1775.

ROBINS, BENJAMIN, *New Principles of Gunnery,* London, 1742. This work represents the beginning of scientific ballistics.

SAINT-REMY, PIERRE SURIREY DE, *Memoires d'Artillerie,* Paris. This work appeared in three principal editions between 1697 and 1747. Each edition was revised to reflect the latest developments in French artillery, including the adoption of the Vallière System in 1732. The earlier editions are in two volumes, the later ones in three. For French artillery this is a work of the first importance.

SCHEEL, M. DE, *Memoires d'Artillerie,* Copenhagen, 1777.

SMITH, CAPTAIN GEORGE, *An Universal Military Dictionary,* London, 1779.

The New Nation, 1784-1835

ADYE, RALPH WILLETT, *The Bombardier and Pocket Gunner,* Boston, 1804. Reprint of an English edition.

DUANE, WILLIAM, *The American Military Library,* 2 vols., Philadelphia, 1809.

DUANE, WILLIAM, *A Military Dictionary,* Philadelphia, 1810.

HOYT, EPAPHRAS, *Practical Instructions for Military Officers,* Greenfield, 1811.

KOSCIUSZKO, THADDEUS A., *Exercises for Garrison and Field Ordnance,* New York, 1812.

KOSCIUSZKO, THADDEUS A., *Maneuvers of Horse Artillery,* New York, 1800.

LALLEMAND, HENRI DOMINIQUE, *Treatise on Artillery,* translated by James Renwick, 2 vols., New York, 1820.

NESMITH, JAMES H., *The Soldier's Manual,* Philadelphia, 1824, reprinted, Philadelphia, 1963.

O'CONNOR, JOHN M., *A Treatise on the Science of War and Fortification,* New York, 1817. Two volumes of text and an atlas of plates.

SCHEEL, M. DE, *A Treatise of Artillery Containing a New System, or the Alterations Made in the French Artillery Since 1765,* translated by Jonathan Williams, Philadelphia, 1800. One volume of text plus an atlas of plates.

SMITH, AMASA, *A Short Compendium of the Duty of Artillerists,* Worcester, Mass., 1800.

STEVENS, WILLIAM, *A System for the Discipline of the Artillery of the United States of America,* New York, 1797.

A System of Artillery Discipline, Boston, 1813.

A System of Exercise and Instruction of Field-Artillery Including Maneuvers for Light or Horse-Artillery, Boston, 1829.

TOUSARD, LOUIS DE, *American Artillerist's Companion,* Philadelphia, 1809-1813. Two volumes of text and an atlas of plates.

The Apex of the Muzzle-loader, 1836-1865

ABBOT, HENRY L., *Siege Artillery in the Campaigns Against Petersburg with Notes on the 15-inch Gun, Professional Papers Corps of Engineers No. 14,* Washington, 1867.

ANDERSON, ROBERT, *Evolutions of Field Batteries of Artillery,* New York, 1860.

ANDERSON, ROBERT, translator, *Instruction for Field Artillery, Horse and Foot,* Philadelphia, 1839.

ANDREWS, R. SNOWDEN, *Mounted Artillery Drill,* Charleston, 1863.

BARNARD, MAJ. J. G., *Notes on Sea-coast Defence,* New York, 1861.

BENTON, CAPT. J. G., *A Course of Instruction in Ordnance and Gunnery,* New York, 1862.

BUCKNER, LIEUT. W. P., *Calculated Tables of Ranges for Navy and Army Guns,* New York, 1865.

BUELL, AUGUSTUS, *The Cannoneer, Washington,* 1890.

FRENCH, WILLIAM H., BARRY, WILLIAM F., and HUNT, HENRY J., *Instructions for Field Artillery,* Philadelphia, 1861.

GIBBON, JOHN, *The Artillerist's Manual,* New York, 1860.

GILMORE, QUINCY A., *Engineer and Artillery Operations against the Defences of Charleston Harbor in 1863,* New York, 1865.

"Heavy Ordnance," *Report of the Joint Committee on the Conduct of the War,* Thirty-sixth Congress, 3 vols., Washington, 1865 (Vol. II, 1-179).

HOLLEY, ALEXANDER L., *A Treatise on Ordnance and Armor,* New York, 1865.

Instruction for Field Artillery, Horse and Foot, Baltimore, 1845.

Instruction for Heavy Artillery, Washington, 1851.

Instruction for Mountain Artillery, Washington, 1851.

Instructions for Heavy Artillery, Washington, 1863.

Instructions for Making Quarterly Returns of Ordnance and Ordnance Stores, Ordnance Memoranda No. 1, Washington, 1863.

KINGSBURY, C. P., *Treatise on Artillery and Infantry,* New York, 1849.

LYNALL, THOMAS, *Rifled Ordnance,* fifth ed., New York, 1864.

MORDECAI, ALFRED, *Artillery for the Land Service of the United States.* One volume of text and a huge atlas of plates, Washington, 1848, 1849.

The Ordnance Manual for the Use of the Officers of the United States Army, Washington, 1841.

The Ordnance Manual for the Use of the Officers of the United States Army, second ed., prepared by Alfred Mordecai, Washington, 1850.

The Ordnance Manual for the Use of the Officers of the United States Army, third ed., Philadelphia, 1861, reprinted 1862.

OWEN, LT. COL. C. H., *The Principle and Practice of Modern Artillery,* London, 1871.

PATTEN, GEORGE, *Artillery Drill,* New York, 1863.

Ranges of Parrott Guns, New York, 1863.

ROBERTS, JOSEPH, *The Hand-book of Artillery for the Service of the United States,* fifth ed., New York, 1863.

SCOFFERN, JOHN, *Projectile Weapons of War and Explosive Compounds,* London, 1858.

SCOTT, COL. H. L., *Military Dictionary,* New York, 1861.

TENNENT, SIR J. EMERSON, *The Story of the Guns,* London, 1864.

WIARD, NORMAN, *Memorial . . . to the Senate and House of Representatives,* second ed., New York, 1863.

WIARD, NORMAN, *Wiard's System of Field Artillery,* New York, 1863.

SECONDARY WORKS

BERKEBILE, DON H., "The 2¾-inch U. S. Howitzer, 1792-1793," *Military Collector & Historian,* XIII, No. 1 (Spring 1961), 1-7.

BIRKHIMER, WILLIAM E., *Historical Sketch of the Artillery of the United States,* Washington, 1884.

CARPENTER, R. G., "The Citadel Gun," *Military Collector & Historian,* XI, No. 1 (Spring 1959), 1-4.

Downey, Fairfax, *The Guns at Gettysburg*, New York, 1958.

Downey, Fairfax, *Sound of the Guns*, New York, 1955.

Egg, Erich, *Der Tiroler Geschutzguss, 1400-1600*, Innsbruck, 1961.

Erwin, Maj. James Q., "Notes on the Coehorn Mortar," *Military Collector & Historian*, XIII, No. 2 (Summer 1961), 35-42.

Evans, Ronald D., "Notes Concerning Wiard's System of Field Artillery," *Military Collector & Historian*, XIX, No. 4 (Winter 1967), 103-108.

Falk, Stanley, "Artillery for the Land Service: The Development of a System," *Military Affairs*, XXVIII, No. 3 (Fall 1964), 94-122.

Ffoulkes, Charles, *The Gun-Founders of England*, Cambridge, 1937.

Gessler, E. A., *Das Schweizerische Geschützwesen zur Zeit des Schwabenkriegs, 1499*, Zurich, 1927.

Gooding, S. James, *An Introduction to British Artillery in North America*, Ottawa, 1965.

Hackley, Frank W., *A Report on Civil War Explosive Ordnance*, Indian Head, Maryland, c. 1960.

Hall, A. R., *Ballistics in the Seventeenth Century*, Cambridge, 1952.

Hazlett, James C., "The Confederate Napoleon Gun," *Military Collector & Historian*, XVI, No. 4 (Winter 1964), 104-110.

Hazlett, James C., "The Federal Napoleon Gun," *Military Collector & Historian*, XV, No. 4 (Winter 1963), 103-108.

Hazlett, James C., "The Napoleon Gun: Its Origin and Introduction into American Service," *Military Collector & Historian*, XV, No. 1 (Spring 1963), 1-5.

Hazlett, James C., "The Napoleon Gun: Markings, Bore Diameters, Weights, and Costs," *Military Collector & Historian*, XVIII, No. 4 (Winter 1966), 109-119.

Hime, Henry W. L., *The Origin of Artillery*, London, 1915.

Jakobsson, Theodor, *Artilleriet Under Karl XII's-Tiden*, Stockholm, 1943.

Jakobsson, Theodor, *Lantmilitar Beväpning och Beklädnad*, Stockholm, 1938.

Kerksis, Sydney C., and Dickey, Thomas S., *Field Artillery Projectiles of the Civil War, 1861-1865*, Atlanta, 1968.

Larter, Col. Harry C., "Materiel of the First American Light Artillery, 1808-1809," *Military Collector & Historian*, IV, No. 3 (September 1952), 53-63.

Larter, Col. Harry C., "On Horses of the U. S. Artillery, Early 1800's," *Military Collector & Historian*, VI, No. 3 (September 1954), 65-67.

Lewis, Emanuel R., "The Ambiguous Columbiads," *Military Affairs*, XXVIII, No. 3 (Fall 1964), 111-122.

Manucy, Albert, *Artillery Through the Ages*, Washington, 1949.

Meyerson, Ake, *Läder-Kannonen från Tidö*, Stockholm, 1938.

Naisawald, L. Van Loan, *Grape and Cannister*, New York, 1960.

Napoleon, Louis, and Col. M. Favé, *Études sur le Passé et l'Avenir de l'Artillerie*, Paris, 6 vols., 1846-1871.

Peterson, Harold L., "Early Cannon Sketches by Charles Willson Peale," *Military Collector & Historian*, I, No. 4 (December 1949), 8, 9.

Peterson, Harold L., *Notes on Ordnance of the American Civil War, 1861-1865*, Washington, 1959.

Peterson, Mendel L., "Ordnance Material Recovered from an Early Seventeenth Century Wreck Site," *Military Collector & Historian*, XIII, No. 3 (Fall 1961), 69-82.

Peterson, Mendel L., "Ordnance Materials Recovered from a Late Sixteenth Century Wreck Site in Bermuda," *Military Collector & Historian*, XIX, No. 1 (Spring 1967), 1-8.

Picard, Commandant breveté Ernest, and Jouan, Lt. Louis, *L'Artillerie Française au XVIIIe Siècle*, Paris, 1906.

Stephen, Walter W., "The Brooke Guns from Selma," *The Alabama Historical Quarterly*, XX, No. 3 (Fall 1958), 461-478.

Tucker, Col. Cary S., "The Early Columbiads," *Military Collector & Historian*, X, No. 2 (Summer 1958), 40-42.

Tucker, Col. Cary S., "Virginia Military Institute Cadet Battery Guns," *Military Collector & Historian*, XIII, No. 4 (Winter 1961), 110-112.

Weller, Jac, "The Artillery of the American Revolution," *Military Collector & Historian*, VIII, No. 3 (Fall 1956), 63-65; VIII, No. 4 (Winter 1956), 97-101.

Weller, Jac, "The Confederate Use of British Cannon," *Civil War History*, III, No. 2 (June 1957), 135-152.

Weller, Jac, "The Field Artillery of the Civil War," *Military Collector & Historian*, V, No. 2 (June 1953), 29-34; No. 3 (September 1953), 65-70; No. 4 (December 1953), 95-97.

Wilson, Lt. A. W., *The Story of the Gun*, London, 1944.

Wise, Jennings C., *The Long Arm of Lee*, 2 vols., Lynchburg, 1915.

Index